Quasiconformal Maps and Teichmüller Theory

Quasiconformal Maps and Teichmüller Theory

A. Fletcher and V. Markovic
Mathematics Institute, University of Warwick, Coventry, CV4 7AL, UK

OXFORD
UNIVERSITY PRESS

This book has been printed digitally and produced in a standard specification in order to ensure its continuing availability

OXFORD
UNIVERSITY PRESS

Great Clarendon Street, Oxford OX2 6DP
United Kingdom

Oxford University Press is a department of the University of Oxford. It furthers the University's objective of excellence in research, scholarship, and education by publishing worldwide.

Oxford is a registered trade mark of Oxford University Press in the UK and in certain other countries

Reprinted 2012

British Library Cataloguing in Publication Data
Data available

Library of Congress Cataloging in Publication Data
Data available

ISBN 978-0-19-856926-8

Preface

These lecture notes are based on a series of lectures entitled *Quasiconformal Maps* given by Vladimir Markovic at the University of Warwick during the spring of 2003. The lectures were typed up and edited by Alastair Fletcher, and it is this which forms the basis of these lecture notes.

The original lectures were intended as a graduate course, and therefore, this book should be accessible to anyone with a grounding in complex analysis, and wants an introduction to the theory of quasiconformal maps and Teichmüller theory. Familiarity with Riemann surfaces and hyperbolic geometry would also be an advantage.

The earliest use of quasiconformal mappings dates back to Gauss. He used quasiconformal mappings to prove that every orientable C^2-surface in $\mathbb{R}^3$ can be made into a Riemann surface. For Gauss, quasiconformal mappings were just a tool in differential geometry, and it was not until Teichmüller that the subject was considered in its own right.

Teichmüller noticed the connection between quasiconformal mappings and function theory, proving (see [54, 55]) that every holomorphic quadratic differential on a compact Riemann surface of genus greater than one determines a quasiconformal mapping which is uniquely extremal in its homotopy class, where extremality here means that the mapping has the smallest deviation from a conformal mapping. Teichmüller showed that all extremal quasiconformal mappings are obtained in this way for compact Riemann surfaces with genus greater than one.

Teichmüller did not prove all of his results, and the theory was put on a sounder footing by Ahlfors in [2]. Ahlfors introduced a more flexible definition for quasiconformal mappings, systematically covering the theory of quasiconformal mappings of the plane.

Bers introduced another approach to Teichmüller theory, again leading to quadratic differentials, but this time as Schwarzian derivatives of conformal extensions of quasiconformal mappings. This approach has the advantage of applying to non-compact Riemann surfaces.

Chapters 1, 2, 3, 5, and 7 contain the standard introductory material to the subject described above. This has been covered previously in texts such as those of Ahlfors [3], Lehto [31], and Lehto–Virtanen [32]. To save multiple references to these texts, we make a note here of where else standard material to the subject is covered.

In these lecture notes, we cover some, but by no means all, modern developments in the field, particularly in infinite dimensional Teichmüller theory. We give an overview of extremal quasiconformal mappings, which was first studied by Grötzsch [25], and in particular, uniquely extremal quasiconformal maps in which the latest, and perhaps final, word is given in [8]. We present the work started by Royden [45] and continued by others in describing the biholomorphic maps of Teichmüller space.

The theory of quasiconformal maps and Teichmüller theory is so large that some selection in which topics to include has to be made, and will undoubtedly reflect the taste of the authors. In the spirit of a series of lectures, we omit some proofs in order to cover more material, but references to where such proofs may be found are always given. Included in the text are a number of examples and exercises. We believe completing these is an excellent way to get a feel for the material and the ideas involved.

The second author would like to express his gratitude to Edgar Reich, Albert Marden and David Epstein for their constant encouragement and many useful conversations over the years. He would also like to his mentor Miodrag Mateljevic. The authors would also like to thank Alvaro Nolla de Celis for help preparing the diagrams for this book.

Contents

1

The Grötzsch Argument

In this chapter, we use Grötzsch's problem of finding a homeomorphism between two rectangles which is closest to a conformal map to motivate the idea of quasiconformality for C^1-diffeomorphisms. We will see that the affine stretching is our first example of a quasiconformal map which is not conformal. We also include a short summary of the basic theory of Riemann surfaces.

1.1 Maps on rectangles

A mapping f on a connected plane domain Ω is conformal, if f is a homeomorphism of Ω and f is holomorphic in Ω. If Ω has a smooth boundary away from finitely many corners, and is simply connected, fix three points A, B, and C on $\partial\Omega$, and fix three points A_1, B_1, and C_1 on $\partial\mathbb{D}$, then there exists a unique $f : \Omega \to \mathbb{D}$ and f extends continuously to $\partial\Omega$ with $f(A) = A_1$, $f(B) = B_1$ and $f(C) = C_1$. This is one formulation of the Riemann mapping theorem.

Let R be a rectangle with vertices prescribed in clockwise order A, B, C, D and side lengths a and b, and also let R_1 be a rectangle with corresponding vertices A_1, B_1, C_1, D_1 and side lengths a_1 and b_1, and consider $f \in C^1$, where C^1 is the space of continuously differentiable functions, such that $f : R \to R_1$ and $f(A) = A_1$, $f(B) = B_1$ and $f(C) = C_1$.

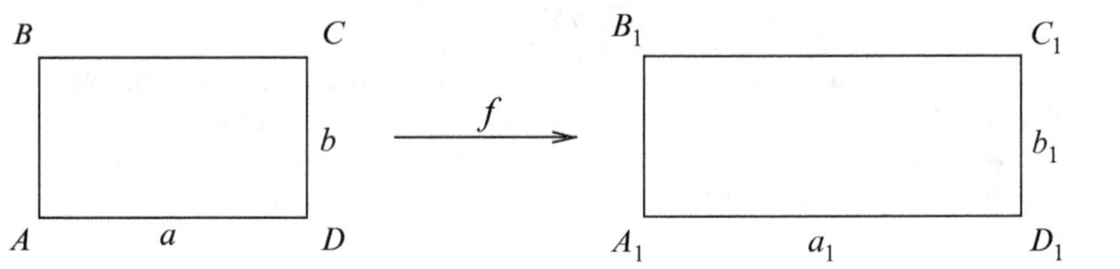

A natural question to ask is whether f can be conformal. One can try the affine map

$$A_f : (x, y) \mapsto \left(\frac{a_1 x}{a}, \frac{b_1 y}{b} \right),$$

which maps R onto R_1, but this is not conformal unless $a/b = a_1/b_1$. In fact, there is only a conformal map from R to R_1 if $a/b = a_1/b_1$.

To see this, assume $f : R \to R_1$ is conformal with the required properties. Then, by the reflection principle, f can be extended to a conformal map from $\mathbb{C}$ to $\mathbb{C}$.

Exercise 1.1.1. Show that this means that f must be of the form $f(z) = \mu z + \nu$ for some $\mu, \nu \in \mathbb{C}$ and hence that $a/b = a_1/b_1$.

Grötzsch [25] asked, what is the map from R to R_1 closest to a conformal map which satisfies the above properties? Before answering this question, we need to make precise what is meant by being closest to a conformal map.

1.2 Some definitions

Let $f : \Omega \to f(\Omega)$ be a C^1-diffeomorphism, and $z = x + iy$ is a coordinate on Ω at z_0 and $w = u + iv$ is a coordinate on $f(\Omega)$ at $f(z_0)$. Then f induces a linear mapping between the corresponding tangent spaces, and the differential defines the linear map given by

$$\begin{aligned} \mathrm{d}u &= u_x \mathrm{d}x + u_y \mathrm{d}y, \\ \mathrm{d}v &= v_x \mathrm{d}x + v_y \mathrm{d}y. \end{aligned}$$

This can be summarized in complex form

$$\mathrm{d}w = \mathrm{d}f = f_z \mathrm{d}z + f_{\bar{z}} \mathrm{d}\bar{z},$$

where the complex derivatives are given by

$$\begin{aligned} f_z &= \frac{1}{2}(f_x - i f_y) \\ f_{\bar{z}} &= \frac{1}{2}(f_x + i f_y) \end{aligned}$$

Since f is a diffeomorphism, it is well approximated by a linear map (in this case at z_0). The linear map df maps the unit circle in the z-plane onto an ellipse with major axis of length α and minor axis of length β in the w-plane.

The area of $\mathbb{D}$ is π, and the area of $df(\mathbb{D})$ is $\pi\alpha\beta$, so the Jacobian $J_f(z_0)$ is $\alpha\beta$.

Exercise 1.2.1. Show that $J_f = |f_z|^2 - |f_{\bar{z}}|^2$.

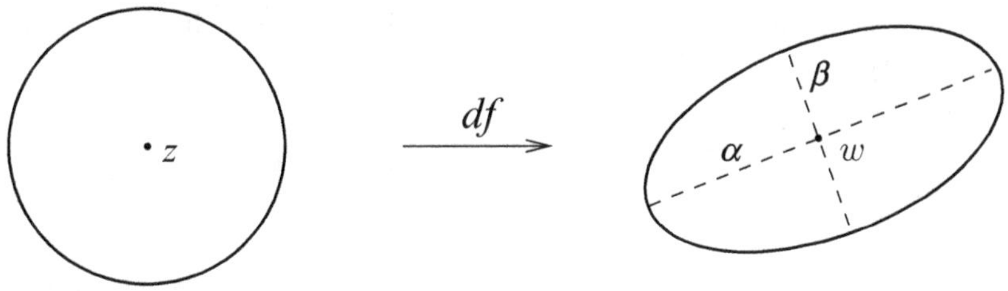

Since sense-preserving maps are being considered here, $|f_z| > |f_{\overline{z}}|$, and so $J_f > 0$. The dilatation (or distortion) at z_0 is defined to be

$$D_f := \frac{\alpha}{\beta} = \frac{|f_z| + |f_{\overline{z}}|}{|f_z| - |f_{\overline{z}}|} \geq 1.$$

The complex dilatation at z_0 is

$$\mu_f = \frac{f_{\overline{z}}}{f_z}.$$

The dilatation and distortion are related by

$$D_f = \frac{1 + |\mu_f|}{1 - |\mu_f|}.$$

By the Cauchy–Riemann equations, if $f \in C^1$, then f is conformal implies that $f_{\overline{z}} \equiv 0$. If f is conformal, $D_f = 1$ and $\alpha = \beta$ above, so df maps circles to circles.

Definition 1.2.1. *Let $f : \Omega \to \mathbb{C}$ be a diffeomorphism. We say that f is a quasiconformal map if $D_f(z)$ is bounded in Ω. We say f is a K-quasiconformal map if $D_f(z) \leq K$ for all $z \in \Omega$. K_f is defined to be the infimum of $K \geq 1$ for which f is K-quasiconformal.*

Every C^1-diffeomorphism is K-quasiconformal on every compactly contained open set in Ω for some K.

Exercise 1.2.2. Show that $K_f = \sup_{z \in \Omega} D_f(z)$. Show that if g and h are conformal mappings, and $\widetilde{f} = h \circ f \circ g$, then $D_f(z) = D_{\widetilde{f}}(g^{-1}(z))$.

A C^1-diffeomorphism is conformal if and only if it is 1-quasiconformal, and so, in some sense, the larger K_f is, the further f is from a conformal map.

1.3 Solving the Grötzsch problem

Now we are in a position to say what is meant by being closest to conformal. Recalling the notation of Section 1.1, we want to know if there is a map f which maps R into R_1, preserving vertices, such that $K_f \leq K_g$ for any other homeomorphism $g : R \to R_1$ with the same properties, and if so, what is K_f?

Exercise 1.3.1. Show that the affine map A_f considered earlier has the following explicit form:

$$A_f(z) = \frac{1}{2}\left(\frac{a_1}{a} + \frac{b_1}{b}\right) z + \frac{1}{2}\left(\frac{a_1}{a} - \frac{b_1}{b}\right)\overline{z}. \tag{1.1}$$

Thus, we get

$$(A_f)_z = \frac{1}{2}\left(\frac{a_1}{a} + \frac{b_1}{b}\right),$$

$$(A_f)_{\overline{z}} = \frac{1}{2}\left(\frac{a_1}{a} - \frac{b_1}{b}\right),$$

and the dilatation of A_f is constant,

$$D_{A_f}(z) = \frac{a_1 b}{b_1 a}.$$

Now consider an arbitrary f with the required properties, and a horizontal path $\gamma \in R$. Then

$$\begin{aligned} a_1 \leq \text{length of } f(\gamma) &= \int_0^a |f_z \mathrm{d}z + f_{\overline{z}} \mathrm{d}\overline{z}| \\ &= \int_0^a |f_z| \left| 1 + \frac{f_{\overline{z}}}{f_z} \right| \mathrm{d}x \end{aligned}$$

because γ is a straight line.

$$\Rightarrow a_1 \leq \int_0^a |f_z||1 + \mu_f| \mathrm{d}x.$$

Now integrate over R.

$$\begin{aligned} a_1 b &\leq \int\int_R |f_z||1+\mu_f| \mathrm{d}x\, \mathrm{d}y \\ &= \int\int_R \frac{|f_z||1+\mu_f|\sqrt{J_f}}{\sqrt{|f_z|^2 - |f_{\overline{z}}|^2}} \mathrm{d}x\, \mathrm{d}y \\ &= \int\int_R \frac{|1+\mu_f|}{\sqrt{1-|\mu_f|^2}} \sqrt{J_f} \mathrm{d}x\, \mathrm{d}y \\ &\leq \left(\int\int_R \frac{|1+\mu_f|^2}{1-|\mu_f|^2} \mathrm{d}x\, \mathrm{d}y\right)^{1/2} \left(\int\int_R J_f \,\mathrm{d}x\, \mathrm{d}y\right)^{1/2} \end{aligned}$$

using the L^2 version of the Cauchy–Schwarz inequality. Now, since J_f is the area of $f(R)$,

$$\frac{a_1 b^2}{b_1} \leq \int\int_R \frac{|1+\mu_f|^2}{1-|\mu_f|^2} \mathrm{d}x\, \mathrm{d}y \tag{1.2}$$

From (1.2),

$$\frac{a_1 b^2}{b_1} \leq \int\int_R D_f \, dx \, dy$$
$$\Rightarrow \frac{a_1 b}{b_1 a} \leq K_f.$$

So the affine map A_f gives the smallest K_f, i.e., if $f : R \to R_1$, then $K_f \geq K_{A_f}$.

Exercise 1.3.2. Show that the above affine map given by (1.1) is the only such map from R to R_1 which has dilatation $(a_1 b)/(b_1 a)$. (Hint: all the inequalities above must be equalities.)

1.4 Composed mappings

Let $f : \Omega \to f(\Omega)$ and $g : f(\Omega) \to \mathbb{C}$, with z a coordinate on Ω and ζ a coordinate on $f(\Omega)$.

Exercise 1.4.1. Show that

$$\overline{(f_z)} = (\overline{f})_{\overline{z}}$$

and

$$\overline{(f_{\overline{z}})} = (\overline{f})_z.$$

Now show that the complex derivatives of a composed mapping $g \circ f$ are

$$\begin{aligned} (g \circ f)_z &= (g_\zeta \circ f) f_z + (g_{\overline{\zeta}} \circ f) \overline{(f_{\overline{z}})}, \\ (g \circ f)_{\overline{z}} &= (g_\zeta \circ f) f_{\overline{z}} + (g_{\overline{\zeta}} \circ f) \overline{(f_z)}. \end{aligned} \tag{1.3}$$

Solving these equations for $g_\zeta \circ f$ and $g_{\overline{\zeta}} \circ f$ gives

$$\begin{aligned} g_\zeta \circ f &= \frac{1}{J_f} \left((g \circ f)_z \overline{(f_{\overline{z}})} - (g \circ f)_{\overline{z}} \overline{(f_z)} \right) \\ g_{\overline{\zeta}} \circ f &= \frac{1}{J_f} \left((g \circ f)_{\overline{z}} f_z - (g \circ f)_z f_{\overline{z}} \right), \end{aligned} \tag{1.4}$$

where J_f is the Jacobian of f (recall Exercise 1.2.1). Now let $g = f^{-1}$, then from (1.4) we get

$$(f^{-1})_\zeta \circ f = \frac{\overline{(f_z)}}{J_f},$$

$$(f^{-1})_{\overline{\zeta}} \circ f = -\frac{f_{\overline{z}}}{J_f},$$

and so

$$\mu_{f^{-1}} \circ f = \left(-\left(\frac{f_z}{|f_z|}\right)^2 \mu_f\right).$$

From (1.4), we get

$$\mu_{g \circ f} = \frac{\mu_f + r_f(\mu_g \circ f)}{1 + r_f \overline{\mu_f}(\mu_g \circ f)}, \tag{1.5}$$

where $r_f = \overline{f_z}/f_z$. If g is conformal,

$$\mu_{g \circ f} = \mu_f, \quad \text{and} \quad D_g \circ f = D_f.$$

If f is conformal,

$$\mu_{g \circ f} = (\mu_{g \circ f}) \frac{\overline{(f_z)}}{f_z}, \quad \text{and} \quad D_{g \circ f} = D_g.$$

Finally, if $h = g \circ f$,

$$\mu_{h \circ f^{-1}} \circ f = \left(\frac{f_z}{\overline{(f_z)}}\right) \frac{\mu_h - \mu_f}{1 - (\overline{\mu_f})\mu_h}. \tag{1.6}$$

Theorem 1.4.1. *If f is K-quasiconformal, and g is K_1-quasiconformal, then $f \circ g$ and $g \circ f$ are both KK_1-quasiconformal maps.*

Proof. As a shorthand, g_ς will here be used to denote $g_\varsigma \circ f$. Now the dilatation of $g \circ f$ is given by

$$\begin{aligned} D_{g \circ f} &= \frac{|(g \circ f)_z| + |(g \circ f)_{\bar{z}}|}{|(g \circ f)_z| - |(g \circ f)_{\bar{z}}|} \\ &= \frac{|g_\varsigma f_z + g_{\bar{\varsigma}} \overline{f_{\bar{z}}}| + |g_\varsigma f_{\bar{z}} + g_{\bar{\varsigma}} \overline{f_z}|}{|g_\varsigma f_z + g_{\bar{\varsigma}} \overline{f_{\bar{z}}}| - |g_\varsigma f_{\bar{z}} + g_{\bar{\varsigma}} \overline{f_z}|} \\ &\le \frac{|f_z||g_\varsigma| + |f_z||g_{\bar{\varsigma}}| + |f_{\bar{z}}||g_\varsigma| + |f_{\bar{z}}||g_{\bar{\varsigma}}|}{|f_z||g_\varsigma| + |f_z||g_{\bar{\varsigma}}| - |f_{\bar{z}}||g_\varsigma| - |f_{\bar{z}}||g_{\bar{\varsigma}}|} \\ &= D_f D_g. \end{aligned}$$

The calculation runs analogously for $f \circ g$. □

Example 1.4.1. Let z be given by $re^{i\theta}$ in polar coordinates, and consider the mapping $w = f(z) = r^t e^{i\theta}$, where $t > 0$ is a fixed parameter. Despite the fact that f is not bi-Lipschitz near 0, f is actually a quasiconformal mapping of $\mathbb{C}$. We see that

$$\log w = t \log r + i\theta$$

and so, locally, f is of the form conformal∘affine∘conformal, where the affine mapping is a stretching by a factor t. Therefore f is a K-quasiconformal mapping of $\{0 < |z| < \infty\}$ onto $\{0 < |w| < \infty\}$, where $K = \max\{t, 1/t\}$. Since $z = 0$ is an isolated singularity, we will see later that it is removable, and so f is a K-quasiconformal mapping of $\mathbb{C}$ onto itself. The complex dilatation is, for $z \neq 0$,

$$\mu_f(z) = \frac{t-1}{t+1}\frac{z}{\bar{z}}$$

and so

$$\sup_{z\in\mathbb{C}} |\mu_f(z)| = \left|\frac{t-1}{t+1}\right| .$$

1.5 Riemann surfaces

In this section, we present some of the basic facts about Riemann surfaces. A real n-dimensional manifold M is a Hausdorff space with a countable base for topology which is locally homeomorphic to the standard euclidean space $\mathbb{R}^n$. In other words, for all $p \in M$, there exists an open neighbourhood $U_p \subset M$ of p and a homeomorphism π_p from U_p onto an open set in $\mathbb{R}^n$. The pair (U_p, π_p) is called a coordinate chart at p. If two coordinate charts (U_p, π_p) and (U_q, π_q) overlap, so that $U_p \cap U_q$ is non-empty, then the map $\pi_q \circ \pi_p^{-1}$ defined on $\pi_p(U_p \cap U_q)$ from an open set of $\mathbb{R}^n$ into $\mathbb{R}^n$ must be a homeomorphism. If there is a collection of charts $\{(U_{p_i}, \pi_{p_i})\}$ for which the union of their domains is the whole of M, then the collection is called an atlas on M.

The complex space $\mathbb{C}^n$ of dimension n can be viewed as the real space $\mathbb{R}^{2n}$ of dimension $2n$ via the identification

$$(x_1 + iy_1, \dots, x_n + iy_n) \longleftrightarrow (x_1, \dots, x_n, y_1, \dots, y_n)$$

and so a complex manifold of dimension n can be viewed as a real manifold of dimension $2n$. A complex manifold is called analytic if all the transition maps $\pi q \circ \pi_p^{-1}$ from $\pi_p(U_p \cap U_q)$ into $\mathbb{C}^n$ are biholomorphic.

Definition 1.5.1. *A Riemann surface is a one-dimensional complex analytic manifold.*

The analytic structure of a Riemann surface is also called a conformal structure, since the transition maps between charts are, in this case, always conformal. Analytic functions can be defined between Riemann surfaces in the following way. A function $f : M \to N$ between Riemann surfaces is analytic at $p \in M$ if there exists coordinate charts π_p on M and $\pi_{f(p)}$ on N such that $\pi_{f(p)} \circ f \circ \pi_p^{-1}$ is holomorphic at $\pi_p(p)$. This is invariant under transition maps between coordinate charts. We then say that $f : M \to N$ is analytic if it is analytic at each $p \in M$.

A collection φ of complex functions φ_i defined on open sets U_i is called an (m, n)-differential on M if

$$\varphi_i \left(\frac{\mathrm{d}\pi_i}{\mathrm{d}\pi_j}\right)^m \left(\overline{\frac{\mathrm{d}\pi_i}{\mathrm{d}\pi_j}}\right)^n = \varphi_j$$

on $U_i \cap U_j$. The differential φ is called holomorphic if each φ_i is holomorphic. Important cases are quadratic differentials, for which $m = 2$ and $n = 0$, and Beltrami differentials, for which $m = -1$ and $n = 1$. We now consider some properties of general surfaces.

Definition 1.5.2. *A smooth covering surface of a surface M, is a pair (S, f) where S is a surface and $f : S \to M$ is a local homeomorphism.*

The map f is a projection. A smooth covering surface (S, f) of M is called regular if every path in M has a lift to S from each point lying over the initial point. In this case, f is surjective.

Theorem 1.5.1 (Monodromy theorem). *Let (S, f) be a regular covering surface of M and γ_0, γ_1 be homotopic paths in M. Then the lifts of γ_0, γ_1 on S from the same initial point have the same end point and are homotopic.*

If a covering surface S is simply connected, then S is called the universal covering surface. The universal covering surface is the maximal covering surface, in the sense that is S' is any other covering surface of M, then S covers S'. Every surface has a universal covering surface.

If (S, f) is a smooth covering surface of a surface M, then a cover transformation g of S over M is a homeomorphism $g : S \to S$ such that $f \circ g = f$. The collection of all such mappings g form a group G under composition, which is called the covering group of S over M. If the projection $f : S \to M$ is surjective and the covering group G of S over M is transitive, then the quotient surface S/G is homeomorphic to M.

Every point of a covering surface S has a neighbourhood in which no two points are equivalent under the action of the covering group, and except for the identity, no covering transformation has any fixed points. The covering group of a universal covering surface S over M is transitive, and so for a universal covering surface, S/G is always homeomorphic to M.

Now let M be a Riemann surface and (S, f) be a smooth covering surface of M. Then S carries a unique conformal structure which makes f analytic. This implies that the covering transformations of S over M are conformal. The following remarkable theorem classifies Riemann surfaces.

Theorem 1.5.2 (Riemann mapping theorem). *Every simply connected Riemann surface is conformally equivalent to exactly one of $\mathbb{D}$, $\mathbb{C}$ or $\widehat{\mathbb{C}} = \mathbb{C} \cup \{\infty\}$.*

Every Riemann surface has exactly one of $\mathbb{D}$, $\mathbb{C}$, $\widehat{\mathbb{C}}$ as its universal covering surface. So given a Riemann surface M and its universal cover S, which is either $\mathbb{D}$, $\mathbb{C}$ or $\widehat{\mathbb{C}}$, and let G be the covering group of S over M, then M is conformally equivalent to S/G. Any Riemann surface which has $\mathbb{D}$ as its universal cover is called hyperbolic. Any non-trivial covering group of $\mathbb{C}$ is generated by either one or two translations, corresponding respectively to an infinite cylinder or a torus, and any Riemann surface with $\widehat{\mathbb{C}}$ as its universal covering surface is exactly $\widehat{\mathbb{C}}$, and so most Riemann surfaces are hyperbolic.

Exercise 1.5.1. Let M be a closed hyperbolic Riemann surface. The automorphism group of M, $\mathrm{Aut}(M)$, is the group of conformal self-mappings of M. Show that $\mathrm{Aut}(M)$ is a finite group.

In fact, Hurwitz proved in [26] that

$$|\mathrm{Aut}(M)| \leq 84(g-1),$$

where g is the genus of M.

2

Geometric definition of quasiconformal maps

The aim of this chapter is to remove the C^1 condition in our definition of quasiconformal maps. The reason for this can be illustrated in the following example.

Example 2.0.1. Consider the homeomorphism $f : (-1, 1) \to (-1, 2)$ given by

$$f(x) = x \quad \text{for} \quad -1 < x < 0 \quad \text{and} \quad f(x) = 2x \quad \text{for } 0 \leq x < 1.$$

This function is not continuously differentiable at $x = 0$, but we can find C^1-homeomorphisms f_n which converge uniformly to f on $(-1, 1)$ by smoothing f on a neighbourhood of 0. Let $\Omega \subset \mathbb{C}$ be the square $(-1, 1) \times (-1, 1)$, and consider the homeomorphisms $g_n : \Omega \to (-1, 2) \times (-1, 2)$ and $g : \Omega \to (-1, 2) \times (-1, 2)$ given by

$$\begin{aligned} g_n(x + iy) &= f_n(x) + i f_n(y), \\ g(x + iy) &= f(x) + i f(y). \end{aligned}$$

Each g_n is C^1 and, by suitable choice of the f_n, is 2-quasiconformal by construction. However, g is 2-quasiconformal, but not C^1, on Ω, even though $g_n \to g$ uniformly on Ω.

This example shows that the C^1 condition on quasiconformal maps turns out to be an unnatural condition since we want to have compactness properties for K-quasiconformal maps. Our aim is to extend the definition of quasiconformal maps so that limits of K-quasiconformal C^1 maps, which are not themselves C^1, can still be K-quasiconformal. The new definition of quasiconformality involves quadrilaterals and we use the method of extremal length to show that our two notions of quasiconformality agree.

2.1 Extremal length

In order to give the definition of quasiconformality that we need, we first need to introduce the notion of extremal length.

Let Γ be a family of rectifiable curves in a domain Ω. A metric $\rho(z)$ is called admissible if ρ is measurable, $\rho(z) \geq 0$ and $A(\rho) \neq 0$ or ∞, where

$$A(\rho) = \int\int_{\Omega} \rho^2(z)\, \mathrm{d}x\, \mathrm{d}y.$$

For $\gamma \in \Gamma$, the length of γ with respect to ρ is

$$L_\gamma(\rho) = \int_\gamma \rho |\mathrm{d}z|$$

and we define

$$L(\rho) = \inf_{\gamma \in \Gamma} L_\gamma(\rho).$$

Definition 2.1.1. *The extremal length of* Γ *is*

$$\lambda(\Gamma) = \sup \frac{L^2(\rho)}{A(\rho)},$$

where the supremum is taken over admissible metrics ρ.

Extremal length can be thought of as an average minimal length for curve families. We say that $\Gamma_1 < \Gamma_2$ if for all $\gamma_2 \in \Gamma_2$, there exists $\gamma_1 \in \Gamma_1$ such that $\gamma_1 \subseteq \gamma_2$.

Lemma 2.1.1. *If* $\Gamma_1 < \Gamma_2$ *then* $\lambda(\Gamma_1) < \lambda(\Gamma_2)$.

Proof. Fix ρ, then $L_{\gamma_1} \leq L_{\gamma_2}$, so the inequality holds if the infimum is taken on both sides. Thus $\lambda(\Gamma_1) \leq \lambda(\Gamma_2)$. □

Example 2.1.1. Let Γ be the family of curves connecting vertical sides of a rectangle R, and each curve does not leave R.

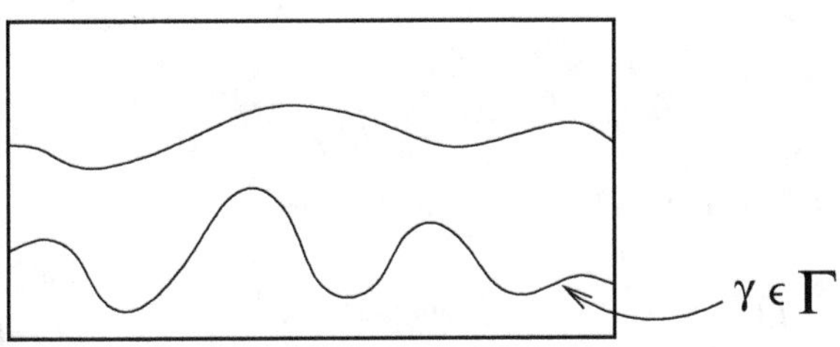

Fix ρ. Then

$$L(\rho) \leq \int_0^a \rho(x + iy)\, \mathrm{d}x,$$

where the integral is over a horizontal curve.

$$(bL(\rho))^2 \le \left(\int\int_R \rho \,\mathrm{d}x\,\mathrm{d}y\right)^2 \le ab \int\int_R \rho^2 \,\mathrm{d}x\,\mathrm{d}y$$

by using the Cauchy–Schwarz inequality. This then gives

$$\frac{L^2(\rho)}{A(\rho)} \le \frac{a}{b}$$

and so

$$\lambda(\Gamma) \le \frac{a}{b}.$$

Take $\rho_0(z) = 1$ for $z \in R$ and $\rho_0(z) = 0$ for $z \notin R$, then

$$\frac{L^2(\rho_0)}{A(\rho_0)} = \frac{a}{b}.$$

Thus $\lambda(\Gamma) = a/b$.

Example 2.1.2. Let $A = \{z : r_1 \le |z| \le r_2\}$, and Γ be the family of curves that separate the two boundary circles, with the curves not leaving A.

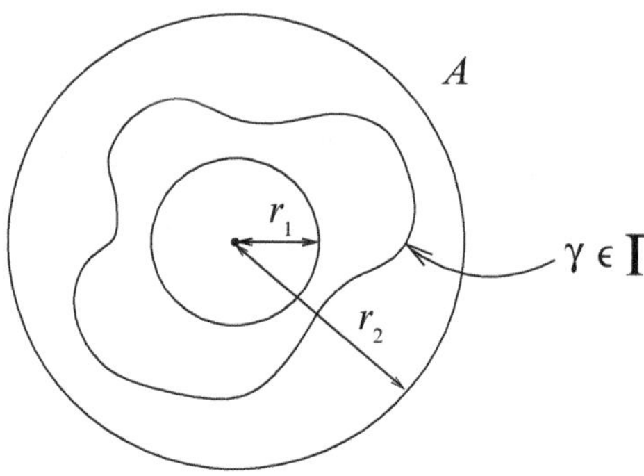

Any path is longer than the circular path following the inner boundary component of A.

$$L(\rho) = \inf_{\gamma\in\Gamma} L_\gamma(\rho) \le \int_0^{2\pi} \rho(re^{i\theta})\,\mathrm{d}\theta$$

for some $r \in [r_1, r_2]$. Then

$$\left(\log\left(\frac{r_2}{r_1}\right) L(\rho)\right)^2 \le \left(\int_{r_1}^{r_2}\int_0^{2\pi} \rho(re^{i\theta})\mathrm{d}\theta\frac{\mathrm{d}r}{r}\right)^2 \le \int_A \rho^2 \int_A \mathrm{d}\theta\,\mathrm{d}r$$

by the Cauchy–Schwarz inequality, and so

$$\frac{L^2(\rho)}{A(\rho)} \le \frac{2\pi}{\log(\frac{r_2}{r_1})}.$$

Take $\rho_0(z) = 1$ for $z \in A$ and 0 for $z \notin A$, then

$$\lambda(\Gamma) = \frac{L^2(\rho_0)}{A(\rho_0)} = \frac{2\pi}{\log(\frac{r_2}{r_1})}.$$

2.2 Curve families

Theorem 2.2.1. *Let $f : \Omega \to f(\Omega)$ be a C^1 K-quasiconformal map, and let Γ be a family of curves which do not leave Ω. Let $\Gamma_1 = f(\Gamma) \subset f(\Omega)$. Then*

$$\frac{\lambda(\Gamma)}{K} \leq \lambda(\Gamma_1) \leq K\lambda(\Gamma).$$

Proof. Let ρ be admissible for Γ and define

$$\rho_1(\zeta) := \left(\frac{\rho}{|f_z| - |f_{\bar{z}}|} \circ f^{-1}\right)(\zeta)$$

where $f(z) = \zeta$ and ρ_1 is admissible for Γ_1. Then

$$\int\int_{f(\Omega)} \rho_1^2 |\mathrm{d}\zeta|^2 = \int\int_{\Omega} \rho^2 \frac{|f_z| + |f_{\bar{z}}|}{|f_z| - |f_{\bar{z}}|} \leq K_f A(\rho).$$

If $\gamma_1 = f(\gamma)$, for $\gamma \in \Gamma$ and $\gamma_1 \in \Gamma_1$, then

$$\int_{\gamma_1} \rho_1 |\mathrm{d}\zeta| \geq \int_{\gamma} \rho |\mathrm{d}z|$$

which implies

$$L_{\gamma_1}(\rho_1) \geq L_{\gamma}(\rho)$$

and hence

$$L(\rho_1) \geq L(\rho)$$

and so, $\lambda(\Gamma_1) \geq K_f^{-1}\lambda(\Gamma)$. Apply the same argument to f^{-1} for the other inequality. □

This theorem shows that extremal length is a conformal invariant (i.e. the case where $K = 1$), which means that a conformal map preserves the extremal length of a family of curves. If Γ_1, Γ_2 are families of curves, then we define

$$\Gamma_1 + \Gamma_2 = \{\gamma_1 + \gamma_2 : \gamma_1 \in \Gamma_1, \gamma_2 \in \Gamma_2, \text{ where this gives a new curve}\}.$$

Theorem 2.2.2. *Let Γ_1, Γ_2 be two disjoint families of curves. Then*

(a) $\lambda(\Gamma_1 + \Gamma_2) \geq \lambda(\Gamma_1) + \lambda(\Gamma_2)$,

(b) $\frac{1}{\lambda(\Gamma_1 \cup \Gamma_2)} \geq \frac{1}{\lambda(\Gamma_1)} + \frac{1}{\lambda(\Gamma_2)}$.

Proof. For (a), assume that $\lambda(\Gamma_1), \lambda(\Gamma_2) \in (0, \infty)$, otherwise there is nothing to prove. Normalize so that $L_1(\rho_1) = A(\rho_1)$ and $L_2(\rho_2) = A(\rho_2)$, and choose

$\rho = \max(\rho_1, \rho_2)$. Then we have

$$L(\rho) \geq L_1(\rho_1) + L_2(\rho_2) = A(\rho_1) + A(\rho_2)$$

and

$$A(\rho) \leq A(\rho_1) + A(\rho_2),$$

and so we get

$$\lambda(\Gamma_1 + \Gamma_2) = \sup \frac{L(\rho)^2}{A(\rho)} \geq A(\rho_1) + A(\rho_2) = \frac{L_1(\rho_1)^2}{A(\rho_1)} + \frac{L_2(\rho_2)^2}{A(\rho_2)}$$

and this implies the inequality in (a).

For (b), if $\lambda(\Gamma_1 \cup \Gamma_2) = 0$, then there is nothing to prove. Otherwise consider an admissible ρ with $L(\rho) > 0$ and set $\rho_1 = \rho$ on E_1, $\rho_2 = \rho$ on E_2 and 0 elsewhere, where E_1 and E_2 are complementary measurable sets satisfying $\Gamma_1 \subseteq E_1$ and $\Gamma_2 \subseteq E_2$. Then we have $L_1(\rho_1) \geq L(\rho)$, $L_2(\rho_2) \geq L(\rho)$, and $A(\rho) = A(\rho_1) + A(\rho_2)$, which gives

$$\frac{A(\rho)}{L(\rho)^2} \geq \frac{A(\rho_1)}{L_1(\rho_1)^2} + \frac{A(\rho_2)}{L_2(\rho_2)^2}$$

and this gives the inequality in (b). □

2.3 Geometric definition of quasiconformal maps

A quadrilateral Q is a Jordan domain with 4 distinguished boundary points. Let f map Q conformally onto a rectangle with sides of length a and b (and recall we can do this as a corollary of the Riemann mapping theorem). If we fix one of the vertices of this rectangle to be 0, and another to be 1, then this rectangle $R(0, 1, 1 + Mi, Mi)$ is called the canonical rectangle of Q.

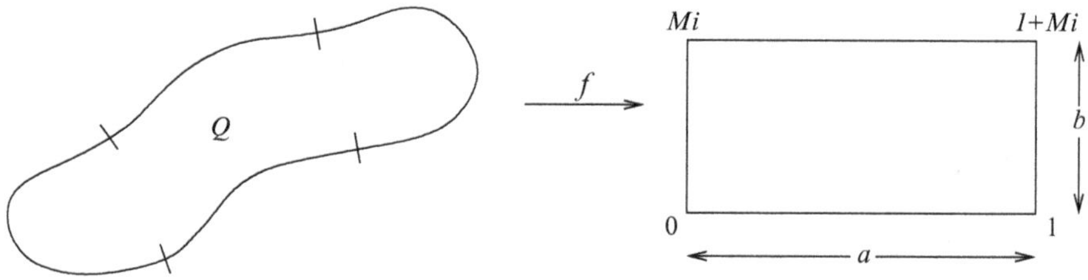

Then the a-sides and b-sides of the original quadrilateral Q can be identified. The quadrilateral $\widetilde{Q}$ is Q with the a and b sides swapped. If Γ is the family of curves connecting the b-sides, then the module of Q is

$$m(Q) = \lambda(\Gamma)^{-1} = \frac{b}{a}.$$

Then $m(\widetilde{Q}) = m(Q)^{-1} = a/b$. Also, using the canonical rectangle given above yields $m(Q) = m(R) = M$. This brings us to our new definition of quasiconformality for homeomorphisms.

Definition 2.3.1. *Let $f : \Omega \to f(\Omega)$ be a homeomorphism. We say that f is K-quasiconformal if for every quadrilateral $Q \subset \Omega$,*

$$\frac{m(Q)}{K} \leq m(f(Q)) \leq Km(Q).$$

The image of a circle under a quasiconformal mapping is called a quasicircle, and the domain bounded by a quasicircle is called a quasidisk. Directly from this definition of quasiconformality we have the following properties:

(1) f is K-quasiconformal $\Rightarrow f^{-1}$ is K-quasiconformal.

(2) If g is conformal and f is K-quasiconformal, then $f \circ g$ and $g \circ f$ are K-quasiconformal.

(3) If g is K_1-quasiconformal and f is K_2-quasiconformal, then $f \circ g$ and $g \circ f$ are $K_1 K_2$-quasiconformal.

Theorem 2.3.1. *If $f : \Omega \to f(\Omega)$ is assumed to be C^1, then this definition of quasiconformality agrees with Definition 1.2.1.*

Proof. First assume that $D_f(z) \leq K$ for all $z \in \Omega$. Let $Q \subset \Omega$ be a quadrilateral, and g be the mapping induced from the canonical rectangle $R(0, 1, 1 + Mi, Mi)$ of Q onto the canonical rectangle $R'(0, 1, 1 + M'i, M'i)$ of $f(Q)$. Then $D_g(z) \leq K$ for $z \in Q$ by Exercise 1.2.2. Therefore

$$|g_x|^2 \leq |f_z|^2 + |f_{\overline{z}}|^2 \leq KJ_g$$

since g is K-quasiconformal and where J_g is the Jacobian of g. Then a length-area argument gives

$$\begin{aligned} M' = m(R') &= \int_R J_g(z)\,\mathrm{d}x\,\mathrm{d}y \geq \frac{1}{K}\int_R |g_x(z)|^2\,\mathrm{d}x\,\mathrm{d}y \\ &\geq \frac{1}{MK}\int_0^1 \mathrm{d}y \left(\int_0^M |g_x(z)|\,\mathrm{d}x\right)^2 \geq \frac{M'^2}{MK}, \end{aligned}$$

and we can conclude that

$$m(f(Q)) = M' \leq KM = Km(Q).$$

For the converse, by composing with suitable conformal mappings, without loss of generality we can look at the point $z = 0$ and assume $f(0) = 0$, $f_z(0) = |f_z(0)|$ and $f_{\overline{z}}(0) = |f_{\overline{z}}(0)|$. Recall that

$$D_f(0) = \frac{|f_z(0)| + |f_{\overline{z}}(0)|}{|f_z(0)| - |f_{\overline{z}}(0)|}.$$

Then

$$f(z) = u(z) + iv(z) = |f_z(0)|z + |f_{\overline{z}}(0)|\overline{z} + o(z).$$

Choose $\delta > 0$ so that the closure of the rectangle

$$R_\delta((-1-i)\delta, (1-i)\delta, (1+i)\delta, (-1+i)\delta)$$

is contained in Ω. Then the b-sides of R'_δ, the image of R_δ under f, are at a mutual distance

$$s_b = 2\delta(|f_z(0)| + |f_{\overline{z}}(0)|) + o(\delta).$$

Now R'_δ has area

$$A(R'_\delta) = 4\delta^2(|f_z(0)|^2 - |f_{\overline{z}}(0)|^2) + o(\delta^2)$$

and Rengel's inequality (see [31]), $m(Q) \leq A(Q)/s_b^2$ for any quadrilateral Q, gives

$$m(R'_\delta) \geq \frac{4\delta^2(|f_z(0)| + |f_{\overline{z}}(0)|)^2 + o(\delta^2)}{4\delta^2(|f_z(0)|^2 - |f_{\overline{z}}(0)|^2) + o(\delta^2)}.$$

Since $m(R_\delta) = 1$, then

$$m(R'_\delta) = \frac{m(R'_\delta)}{m(R_\delta)} \leq K$$

using the hypothesis of the module of a quadrilateral definition of quasiconformality. Then

$$(|f_z(0)| + |f_{\overline{z}}(0)|)^2 + \frac{o(\delta)}{\delta} \leq K\left(|f_z(0)|^2 - |f_{\overline{z}}(0)|^2 + \frac{o(\delta)}{\delta}\right)$$

and letting $\delta \to 0$ give the required result. □

Exercise 2.3.1. Let $f : \Omega \to f(\Omega)$ be a homeomorphism of a plane domain Ω, and for $z \in \Omega$ define

$$\alpha_\epsilon(z) = \max_{|h|=\epsilon} |f(z+h) - f(z)|,$$
$$\beta_\epsilon(z) = \min_{|h|=\epsilon} |f(z+h) - f(z)|$$

and

$$D_f(z) = \limsup_{\epsilon \to 0} \frac{\alpha_\epsilon(z)}{\beta_\epsilon(z)}.$$

Show f is quasiconformal if and only if $D_f(z)$ is bounded over all $z \in \Omega$.

Lemma 2.3.1. *Let R be a rectangle with vertical sides of length* 1 *and horizontal sides of length $m > 1$. Let l be a curve joining the horizontal sides. The curve l splits R into two quadrilaterals Q_1 and Q_2. Then*

$$m(Q_1) + m(Q_2) \leq m(R)$$

with equality if and only if l is a vertical line.

Proof. If we take Γ_1 to be the family of curves from the left-horizontal side of R to l, and Γ_2 the family of curves from l to the right-horizontal side of R, then using Theorem 2.2.2 (a) gives us the inequality. If l is a vertical line, then equality holds.

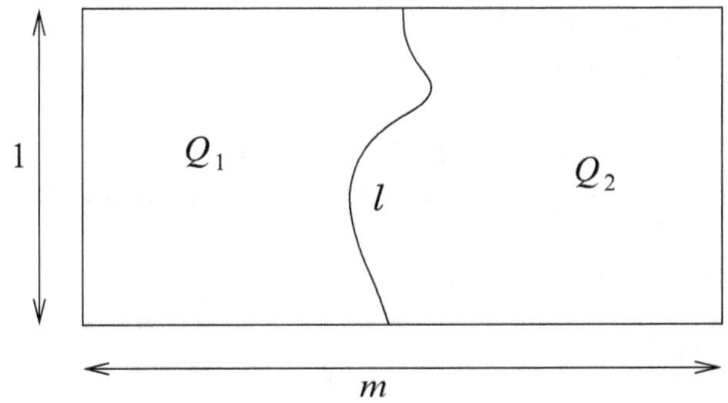

Conversely, let f_i map Q_i conformally onto a rectangle with sides of length 1 and $m_i = m(Q_i)$ for $i = 1, 2$, and we assume $m = m_1 + m_2$. Define a metric ρ on R by $\rho = |f_1'|$ on Q_1, $|f_2'|$ on Q_2 and 0 elsewhere.

$$\begin{aligned}\int\int_R (\rho^2 - 1)\, \mathrm{d}x\, \mathrm{d}y &= \int\int_{Q_1} |f_1'|^2\, \mathrm{d}x\, \mathrm{d}y + \int\int_{Q_2} |f_2'|^2\, \mathrm{d}x\, \mathrm{d}y - \int\int_R \mathrm{d}x\, \mathrm{d}y \\ &= \text{area of } Q_1 + \text{ area of } Q_2 - \text{ area of } R = 0.\end{aligned}$$

Since any horizontal line in R gets mapped onto a curve of length larger than m,

$$\int\int_R (\rho - 1)\, \mathrm{d}x\, \mathrm{d}y \geq 0.$$

Now, we have

$$0 = \int\int_R (\rho^2 - 1)\, \mathrm{d}x\, \mathrm{d}y = \int\int_R (\rho + 1)(\rho - 1)\, \mathrm{d}x\, \mathrm{d}y \geq \int\int_R (\rho - 1)\, \mathrm{d}x\, \mathrm{d}y \geq 0.$$

This implies that $\rho \equiv 1$, or equivalently that f_1 and f_2 are the identity mapping, and the result follows. □

Theorem 2.3.2. *Every 1-quasiconformal map is conformal.*

Proof. Let $f : \Omega \to f(\Omega)$ be 1-quasiconformal, and consider an arbitrary quadrilateral $Q \subset \Omega$. Let f_1 map Q conformally onto a rectangle R with sides of length 1 and m. Let f_2 map the quadrilateral $f(Q)$ conformally onto a rectangle, which since f is 1-quasiconformal, is just a rectangle with the same side lengths as R. Consider the map $g = f_2 \circ f \circ f_1^{-1}$ which maps R onto itself, is 1-quasiconformal, and preserves vertices.

Let l be a vertical line of R, splitting R into two rectangles P_1 and P_2 with horizontal lengths m_1 and m_2 respectively. The map g sends l onto a curve $g(l)$ which splits R into two quadrilaterals Q_1 and Q_2. Now

$$m(Q_1) + m(Q_2) = m(R) = m = m_1 + m_2 = m(P_1) + m(P_2),$$

so $m(P_i) = m(Q_i)$ for $i = 1, 2$. From Lemma 2.3.1, $g(l)$ is vertical. If Q_1 has base length $\tilde{m_1}$, then $\tilde{m_1} = m_1$, so g is the identity on horizontal coordinates. Repeat this argument for vertical coordinates to show g is the identity, and thus $f = f_1 \circ f_2^{-1}$ and is thus conformal on Q, and so conformal on the whole of Ω since Q was arbitrary. □

The power of the extremal length method shows us that a homeomorphism which is 1-quasiconformal must be complex analytic (and in particular, C^1). This need not be the case for K-quasiconformal maps with $K > 1$.

3

Analytic properties of quasiconformal maps

In this chapter, we have a further definition of quasiconformality, which this time is analytic in flavour, and yet agrees with our previous definitions. This definition makes possible a way to show that quasiconformal maps are Hölder continuous and thus satisfy important compactness properties.

3.1 Analytic definition and corollaries

A map $g : (a, b) \to \mathbb{C}$ is absolutely continuous on the interval (a, b) if

$$g(x) = \int_a^x h(t)\, \mathrm{d}t + g(a)$$

for $h \in L^1_{\mathrm{loc}}(a, b)$, the space of locally integrable functions. If g is absolutely continuous, then it is differentiable almost everywhere and $g' = h$ almost everywhere.

Let Ω be a plane domain, $f : \Omega \to f(\Omega) \subset \mathbb{C}$ and let a rectangle $R \subset \Omega$ have sides parallel to the x and y axes. We say f is absolutely continuous on lines (ACL) on R if f is absolutely continuous on almost every horizontal and vertical line in R. The map f is ACL on Ω if f is ACL on every rectangle $R \subset \Omega$. We state several important theorems.

Theorem 3.1.1 (Analytic definition of quasiconformal maps). *A homeomorphism $f : \Omega \to f(\Omega)$ is K-quasiconformal if and only if the following holds:*

(a) *f is ACL on Ω,*

(b) *$|f_{\bar{z}}| \le k|f_z|$ almost everywhere in Ω, where $k = (K-1)/(K+1)$.*

Quasiconformality can also be defined for maps between Riemann surfaces. The map $f : M \to N$ is said to be K-quasiconformal at $p \in M$ if there are coordinate charts (U_p, π_p) on M and $(U_{f(p)}, \pi_{f(p)})$ on N such that $\pi_{f(p)} \circ f \circ \pi_p^{-1}$ is a K-quasiconformal mapping of the plane domain $\pi_p(U_p)$. The mapping f is then said to be K-quasiconformal if it is K-quasiconformal at all $p \in M$. Note that this definition is invariant of the choice of coordinate chart, since the transition maps are conformal.

The proof that the analytic definition of quasiconformality is equivalent to the one given in Section 2.3 relies on the following theorem.

Theorem 3.1.2. *If $f : \Omega \to f(\Omega)$ is a homeomorphism and f_z, $f_{\bar{z}}$ exist almost everywhere in Ω, then f is differentiable almost everywhere in Ω.*

This theorem is not true in higher dimensions, although quasiconformal maps in dimensions greater than or equal to 2 are almost everywhere differentiable (see [32]). The following theorem is a corollary to Theorem 3.1.1.

Theorem 3.1.3. *Let $f : \Omega \to f(\Omega)$ be a homeomorphism and let $\gamma \subset \Omega$ be an analytic Jordan curve. If f is K-quasiconformal on $\Omega \setminus \gamma$, then f is K-quasiconformal on the whole of Ω.*

Remark. In fact, γ can be assumed to be C^1 in the hypothesis of this theorem.

Proof. Let γ be parametrized by $t \in [0,1]$, and then

$$\gamma(t) = \gamma_1(t) + i\gamma_2(t),$$

where γ_1, γ_2 are real analytic functions on $[0,1]$. There exists an open set U such that $(0,1) \subset U \subset \mathbb{C}$ and there exist functions $\gamma_1(z), \gamma_2(z) : U \to \mathbb{C}$ which are holomorphic and are extensions of the real analytic functions. Define

$$\gamma(z) = \gamma_1(z) + i\gamma_2(z),$$

which is holomorphic on U.

Away from the zeroes of $\gamma'(z)$, $\gamma(z)$ is locally univalent. Since γ is an analytic closed Jordan arc, there are only finitely many zeroes of $\gamma'(t)$ on $[0,1]$. Now if $\gamma'(x_0) \neq 0$, take a neighbourhood $\mathbb{D}_{x_0}$ of x_0 with radius r_0. Then γ is holomorphic on $\mathbb{D}_{x_0}$ and $f \circ \gamma : \mathbb{D}_{x_0} \hookrightarrow f(\Omega)$ for small enough r_0.

Thus on the set $\mathbb{D}_{x_0} \setminus [x_0 - r_0, x_0 + r_0]$, $f \circ \gamma$ is a homeomorphism and K-quasiconformal. The composition $f \circ \gamma$ is ACL on almost every horizontal line contained in $\mathbb{D}_{x_0}$. For vertical lines, it is ACL above the segment $[x_0 - r_0, x_0 + r_0]$ and below it, so it is absolutely continuous on the whole of the vertical line (this is true because the derivative of $f \circ \gamma$ does not vanish in $\mathbb{D}_{x_0}$).

So $f \circ \gamma$ is K-quasiconformal on $\mathbb{D}_{x_0}$, which means f is K-quasiconformal on Ω except possibly at a finite set of points where $\gamma'(t) = 0$, but single points are removable for quasiconformality, so f is K-quasiconformal on Ω. □

Another important result is the following.

Theorem 3.1.4. *If $f : \Omega \to f(\Omega)$ is a quasiconformal map, then*

$$|\nabla f| = |f_z| + |f_{\bar{z}}| \in L^2_{\text{loc}}(\Omega).$$

Indeed, it is true that $|\nabla f| \in L^p_{\text{loc}}(\Omega)$ for $0 < p \leq p_0$ for some $p_0 > 2$, where p_0 depends on K.

Exercise 3.1.1. For $K > 1$, define $f : \mathbb{D} \to \mathbb{D}$ by

$$f(z) = f(re^{i\theta}) = r^K e^{i\theta}.$$

Note that away from 0, f is analytic. Show that f is K-quasiconformal and find $p_0(K) > 2$ such that $|\nabla f| \in L^p$ for $p = p_0(K)$, but not for any $p > p_0(K)$.

Theorem 3.1.5. *If a homeomorphism f is ACL and $f_{\bar{z}} = 0$ almost everywhere then f is conformal.*

As a result of this theorem and Theorem 3.1.1, we have the following corollary.

Corollary 3.1.1. *If f is quasiconformal and $f_{\bar{z}} = 0$ almost everywhere, then f is conformal.*

Example 3.1.1. Let C be the Cantor set on $(0, 1)$, then every $x \in C$ can be written as

$$x = \sum_{i=1}^{\infty} 2 \times 3^{-n_i}$$

for some subset $\{n_i\}$ of $\mathbb{Z}^+$. The Cantor function $F : (0, 1) \to (0, 1)$ is defined by setting

$$F(x) = \sum_{i=1}^{\infty} 2^{-n_i}$$

for $x \in C$ and extended to the whole of $(0, 1)$ be requiring that F be monotonically increasing. Since F is constant on connected components of the complement of C, F is differentiable almost everywhere, with derivative 0, but is not differentiable at points of C. Now define the function $f : (0, 1) \times (-\infty, \infty) \to \mathbb{C}$ given by

$$f(x + iy) = F(x) + x + iy.$$

Now, $F(x) + x$ is a homeomorphism of $(0, 1)$ onto $(0, 2)$, and so f is a homeomorphism, which is differentiable almost everywhere, and in fact $f_{\bar{z}} = 0$ almost everywhere. However f cannot be conformal on $(0, 1) \times (-\infty, \infty)$ because it is not differentiable at any point of $C \times (-\infty, \infty)$. Moreover, f cannot be quasiconformal, because if it was, then it would have to be 1-quasiconformal and hence conformal. In conclusion, if $f : \Omega \to f(\Omega)$ is a homeomorphism, differentiable almost everywhere and $f_{\bar{z}} = 0$ almost everywhere, then this does not imply that f is conformal or quasiconformal, and so the ACL condition is certainly necessary in the definition of quasiconformality.

Recall Stokes theorem, which says that if M is a Riemann surface, and ω a compactly supported differential 1-form on M which is of class C^1, then

$$\int_M \mathrm{d}\omega = \int_{\partial M} \omega, \tag{3.1}$$

where ∂M is the boundary of M with its induced orientation. Stokes theorem can be stated in various ways, including the complex version

$$\int_{\Omega} f_{\bar{z}}\, dx\, dy = \int_{\partial\Omega} f\, dz,$$

where Ω is a plane domain.

Theorem 3.1.6. *Let $f : \Omega \to f(\Omega)$ be quasiconformal. Then f maps sets of measure zero onto sets of measure zero. Also, if $E \subset \Omega$ is measurable, then*

$$Area(f(E)) = \int\int_E J\, dx\, dy,$$

where J is the Jacobian of f.

Proof. The map $f = u + iv$ can be approximated by C^2 functions $u_n + iv_n$ such that $u_n \to u$, $v_n \to v$,

$$\int_{\Omega} |u_x - (u_n)_x|^2\, dx\, dy \to 0,$$

and

$$\int_{\Omega} |v_x - (v_n)_x|^2\, dx\, dy \to 0.$$

Consider rectangles R such that u and v are absolutely continuous on all sides of R. By Stokes theorem, (3.1),

$$\int_R ((u_m)_x(v_n)_y - (u_m)_y(v_n)_x)\, dx\, dy = \int_{\partial R} u_m\, dv_n.$$

As $m, n \to \infty$, the left-hand side converges to

$$\int_R J\, dx\, dy.$$

As $m \to \infty$, the right-hand side converges to

$$\int_{\partial R} u\, dv_n = -\int_{\partial R} v_n\, du.$$

Then as $n \to \infty$, this converges to

$$-\int_{\partial R} v\, du = \int_{\partial R} u\, dv.$$

These all converge as claimed because u and v are absolutely continuous on ∂R. Therefore,

$$\int_R J\, dx\, dy = \int_{\partial R} u\, dv$$

and for these rectangles R, the line integral represents the image area. This also shows that f maps sets of measure zero onto sets of measure zero. □

3.2 Extremal ring domains

In order to prove compactness properties of quasiconformal maps, we introduce some ring domains.

(i) This ring domain is due to Grötszh. Let $R_1 \subseteq \mathbb{C}$ be any ring domain separating $\mathbb{D}$ and a point $R \in \mathbb{R}$, and let Γ be the family of curves contained in R_1 and separating the boundary components of R_1.

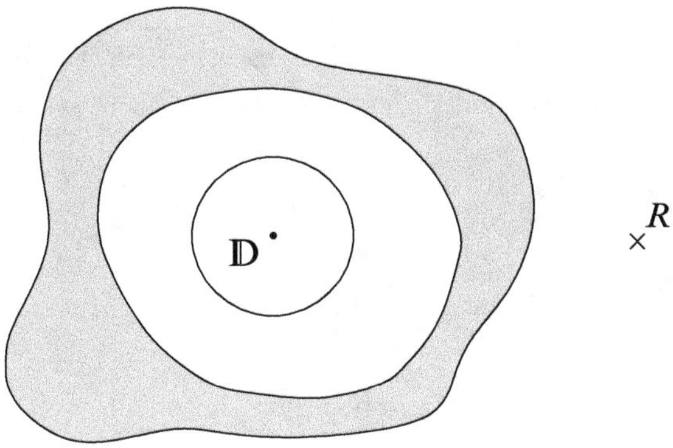

The module is given by $m(R_1) = \lambda(\Gamma)^{-1}$, and

$$\sup m(R_1) = \frac{1}{2\pi} \log \varphi(R)$$

where the supremum is taken over all such ring domains descibed above. The ring domain achieving the supremum is given by $\mathbb{C} \setminus (\overline{\mathbb{D}} \cup [R, \infty))$.

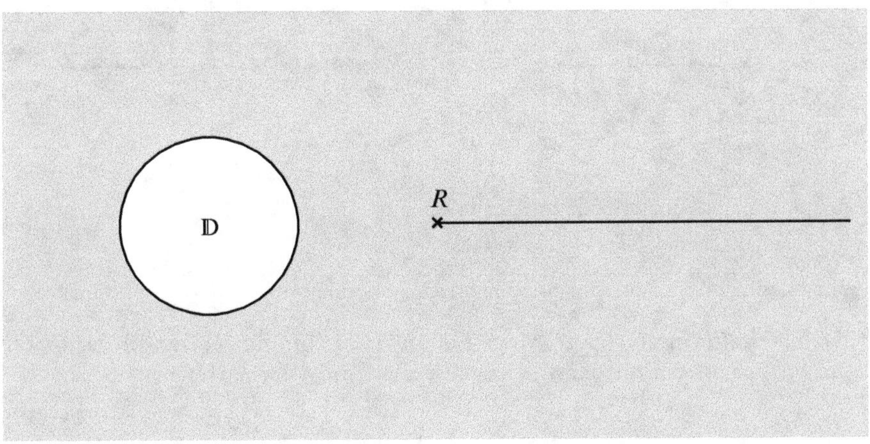

By reflecting in $\partial\mathbb{D}$, we see that the extremal ring domain separating $\partial\mathbb{D}$ and the line segment $[0, z]$ for $z \in \mathbb{D}$ is $\mathbb{D} \setminus [0, z]$, and the module of this ring domain will be denoted by $\mu(|z|)$.

(ii) Teichmüller considered this ring domain. Let the ring domain $R_2 \subseteq \mathbb{C}$ separate $\{0, -1\}$ from $P \in \mathbb{R}^+$, and let Γ be the family of curves separating the boundary components of R_2.

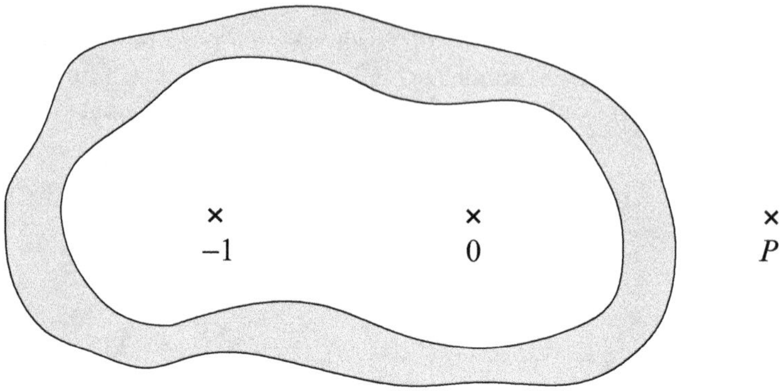

$$\sup m(R_2) = \frac{1}{2\pi} \log \psi(P),$$

and the supremum is achieved by the ring domain $\mathbb{C} \setminus ([-1, 0] \cup [P, \infty))$.

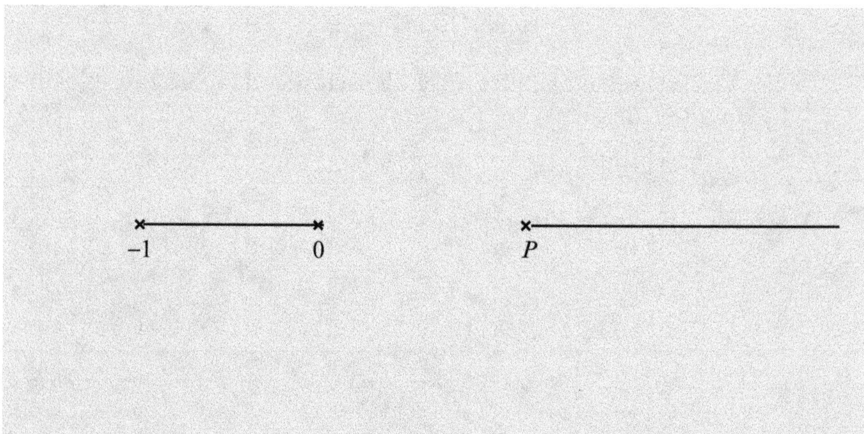

(iii) The following is due to Mori. Let the ring domain R_3 separate the points z_1, $z_2 \in \mathbb{D}$ from 0, and z_1, z_2 are a distance λ apart.

$$\sup m(R_3) = \frac{1}{2\pi} \log X(\lambda),$$

and the extremal domain is given by the following: let A be the arc of $\mathbb{T}$ with z_1, $z_2 \in \mathbb{T}$ and being a distance λ apart, and let B be the one-sided ray from

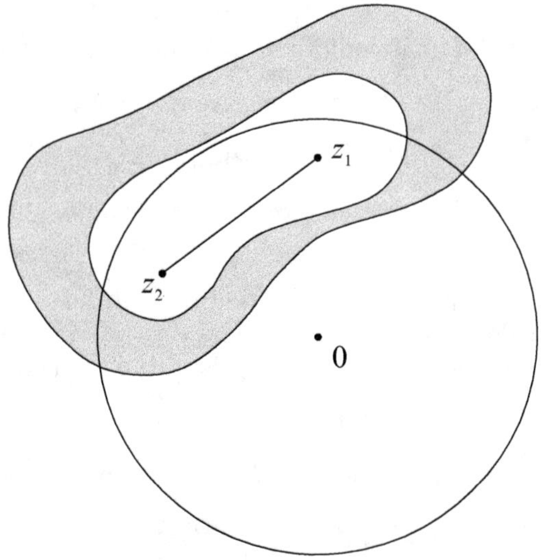

the origin, pointing away from the mid-point of A. Then the extremal domain is given by $\mathbb{C} \setminus (A \cup B)$.

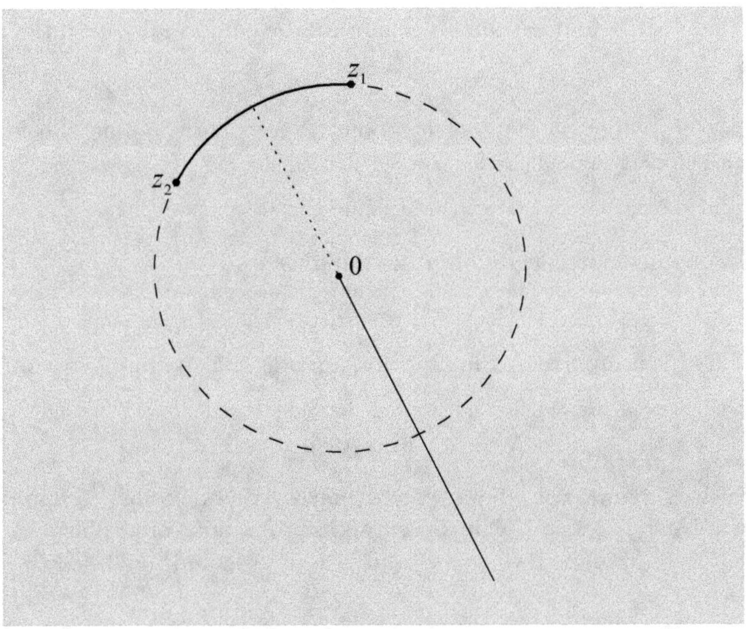

Using a method involving elliptic functions (see [3]), we have the following inequalities

$$\varphi(R) \leq 4R, \tag{3.2}$$

$$\lambda X(\lambda) \leq 16. \tag{3.3}$$

We now prove the Hersch–Pfluger theorem, which uses ring domains to give a bound on the growth of quasiconformal mappings on $\mathbb{D}$ which fix 0.

Theorem 3.2.1. *Let $f : \mathbb{D} \to \mathbb{D}$ be K-quasiconformal and $f(0) = 0$. Then*

$$|f(z)| \leq \varphi_K(|z|)$$

where $\varphi_K : [0, 1] \to \mathbb{R}^+$ is a strictly increasing function. The bound can be attained.

Proof. For a given $z \in \mathbb{D}$, let R be the ring domain formed by $\mathbb{D} \setminus [0, z]$. The module of this ring domain is given by $\mu(R) = \mu(|z|)$ from the Grötzsch ring domain discussion. The image of R under f is a ring domain with module

$$\mu(f(R)) \leq \mu(|f(z)|)$$

since f fixes 0 and ring domains such as R are extremal. Now define

$$\varphi_K(r) = \mu^{-1}\left(\frac{\mu(r)}{K}\right),$$

and then

$$|f(z)| \leq \varphi_K(|z|).$$

To show that this bound can be attained, map R conformally onto the annulus

$$\{w : \mathrm{e}^{-\mu(r)} < |w| < 1\}$$

by the map f_1 which is symmetric about $\mathbb{R}$ and corresponds the boundary components $|z| = 1$ and $|w| = 1$. Now

$$f_2(w) = w|w|^{1/K-1}$$

maps $f_1(R)$ homeomorphically onto the annulus

$$\{t : \mathrm{e}^{-\mu(r)/K} < |t| < 1\},$$

and f_2 is K-quasiconformal. Finally, this annulus can be mapped conformally onto

$$\mathbb{D} \setminus [0, \varphi_K(r)]$$

by f_3, which is symmetric about $\mathbb{R}$ and makes the boundary components $\partial\mathbb{D}$ correspond. Now $f_3 \circ f_2 \circ f_1$ can be extended to a homeomorphism of $\mathbb{D}$ onto itself which fixes 0, maps r to $\varphi_K(r)$ and is K-quasiconformal on $\mathbb{D}$. □

3.3 Hölder continuity

We now show some finer analytic properties of quasiconformal maps, specifically that they are Hölder continuous and satisfy compactness properties as a consequence of this.

Let $f : \overline{\mathbb{D}} \to \overline{\mathbb{D}}$ be a surjective K-quasiconformal map and $f(0) = 0$. By reflecting in $\partial\mathbb{D}$, one can extend f to a map from $\mathbb{C}$ to itself, and the extension is K-quasiconformal.

Theorem 3.3.1. *Let $f : \mathbb{D} \to \mathbb{D}$ be a surjective K-quasiconformal map normalized by $f(0) = 0$. Then for all $z_1, z_2 \in \mathbb{D}$,*

$$|f(z_1) - f(z_2)| \leq 16|z_1 - z_2|^{1/K} \tag{3.4}$$

i.e. f is $1/K$-Hölder continuous.

Proof. Initially assume that $f : \overline{\mathbb{D}} \to \overline{\mathbb{D}}$. This restriction will be removed later. Assume $|z_1 - z_2| < 1$, otherwise the result is trivial. Form the circle which has diameter $[z_1, z_2]$, and from the centre of this circle, consider the circle with radius $1/2$. The region between these two circles forms an annulus A.

Case (i) $A \subset \mathbb{D}$.

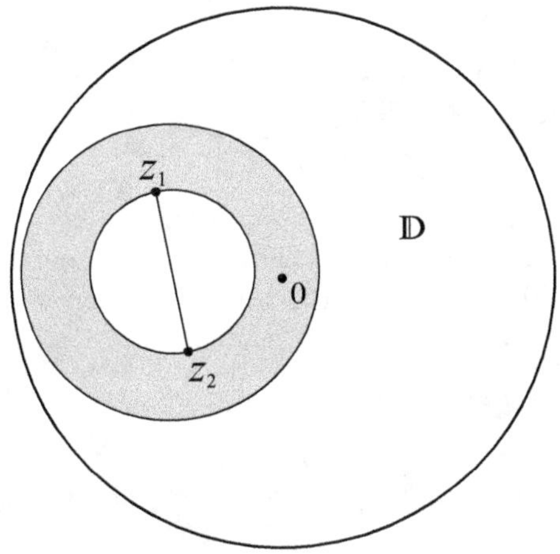

Let f map z_i onto ζ_i for $i = 1, 2$. Compose f with the Mobius transformation T which maps ζ_1 to 0. Now compose with a reflection in $\partial\mathbb{D}$ given by

$$x^* = \frac{1}{\overline{x}},$$

for $x \in \mathbb{D}$. Let $\widetilde{A}$ be the image of A under this composition. $\widetilde{A}$ separates $\mathbb{D}$ from $(T(\zeta_2))^*$. So using (3.2),

$$\frac{1}{2\pi} \log \frac{1}{|z_1 - z_2|} = m(A) \leq Km(\widetilde{A}) \leq \frac{K}{2\pi} \log \varphi(|T(\zeta_2)^*|) \leq \frac{K}{2\pi} \log \frac{8}{|\zeta_1 - \zeta_2|}$$

$$\Rightarrow |f(z_1) - f(z_2)| \leq 16|z_1 - z_2|^{1/K}$$

Case (ii) A does not contain the origin.

The inner boundary component of $\widetilde{A} = f(A)$ intersects $|\zeta| < 1$ in a set with diameter at least $|\zeta_1 - \zeta_2|$. The outer boundary component of $\widetilde{A}$ has the origin in its interior, and so, using (3.3),

$$\frac{1}{2\pi}\log\frac{1}{|z_2-z_1|}\leq\frac{K}{2\pi}\log X(|\zeta_2-\zeta_1|)<\frac{K}{2\pi}\log\frac{16}{|\zeta_2-\zeta_1|}.$$

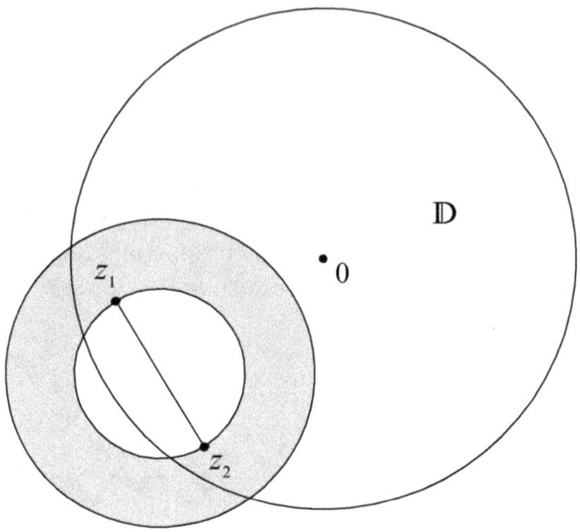

To remove the hypothesis that f is continuous on $|z|=1$, consider the image region of $\mathbb{D}_r=\{z:|z|<r\}$, and map it conformally onto $|w|<1$ by the Riemann map ψ_r satisfying $\psi_r(0)=0$ and $\psi_r'(0)>0$. The composition $\psi_r(f(rz))$ is K-quasiconformal and continuous on $|z|=1$, and so

$$|\psi_r(f(rz_2))-\psi_r(f(rz_1))|<16|z_2-z_1|^{1/K}.$$

Let $r\to 1$, then $\psi_r(\mathbb{D})\to\mathbb{D}$ in the Hausdorff metric. The Carathéodory convergence theorem then implies that ψ_r converges to the identity. This implies

$$|f(z_1)-f(z_2)|\leq 16|z_1-z_2|^{1/K}.$$

Exercise 3.3.1. Use the Hölder continuity of f and f^{-1} to show that f extends continuously to $\partial\mathbb{D}$ and therefore to a homeomorphism of $\overline{\mathbb{D}}$.

This completes the proof. □

Remark. The best constant that can be achieved in Theorem 3.3.1 is 16. If we have two Mori ring domains, $G(\lambda_1)$ and $G(\lambda_2)$, they can be mapped onto annuli and then these two annuli can be mapped to each other by stretching the radii in constant ratio

$$K=\frac{\log X(\lambda_1)}{\log X(\lambda_2)}>1$$

for $\lambda_1<\lambda_2$. Hence a K-quasiconformal map is obtained which makes λ_1 correspond with λ_2, and so

$$\frac{16-\epsilon}{\lambda_2}\leq X(\lambda_2)=X(\lambda_1)^{1/K}\leq\left(\frac{16}{\lambda_1}\right)^{1/K}$$

for small enough λ_2. Thus

$$\lambda_2 \geq (\lambda_1)^{1/K} \frac{16 - \epsilon}{16^{1/K}}$$

and for large enough K the constant factor is arbitrarily close to 16.

3.4 Compactness properties of quasiconformal maps

As a corollary to Theorem 3.3.1, we have the following result on families of quasiconformal maps.

Theorem 3.4.1. *The family of all K-quasiconformal maps $f : \mathbb{D} \to \mathbb{D}$ which are surjective and satisfy $f(0) = 0$, is compact.*

Let d_h denote the hyperbolic distance on $\mathbb{D}$ given by

$$d_h(z_1, z_2) = \tanh^{-1} \left| \frac{z_1 - z_2}{1 - z_1 \overline{z_2}} \right|$$

for $z_1, z_2 \in \mathbb{D}$ (see [4]).

Theorem 3.4.2. *Let $f : \overline{\mathbb{D}} \to \overline{\mathbb{D}}$ be any K-quasiconformal map which fixes* 1, -1 *and i. Then there exists a constant $C(K)$ such that*

$$d_h(0, f(0)) \leq C(K).$$

Proof. The proof is based on the Hölder continuity of quasiconformal maps. Let $T : \mathbb{D} \to \mathbb{D}$ be a Möbius transformation such that $T(f(0)) = 0$. Then $g = T \circ f$ satisfies the assumptions of Theorem 3.3.1, and g is K-quasiconformal. Applying Theorem 3.3.1 to g or its inverse,

$$|1 - (-1)| \leq 16 |T(1) - T(-1)|^{1/K},$$

$$|1 - i| \leq 16 |T(1) - T(i)|^{1/K}.$$

Exercise 3.4.1. Show that this implies there is a constant $C(K)$ such that

$$d_h(T^{-1}(0), 0) \leq C(K).$$

This, then gives the desired result that $d_h(0, f(0)) \leq C(K)$. □

Theorem 3.4.3. *Let $f : \overline{\mathbb{D}} \to \overline{\mathbb{D}}$ be such that f is the identity on $\partial\mathbb{D}$ and f is K-quasiconformal. Then there is a constant $D(K)$ such that*

$$d_h(z, f(z)) \leq D(K)$$

for all $z \in D$.

Exercise 3.4.2. Using the proof of Theorem 3.4.2 as an outline, prove this theorem.

Let $f : S^2 \to S^2$ be K-quasiconformal and fixes 0, 1 and ∞. We want to measure the distance between z and $f(z)$. The thrice punctured sphere

$$S^2 \setminus \{0, 1, \infty\}$$

carries the unique hyperbolic metric. Let

$$\pi : \mathbb{D} \to S^2 \setminus \{0, 1, \infty\}$$

be the covering map. Incidentally, the corresponding group Γ for which

$$\mathbb{D}/\Gamma \simeq S^2 \setminus \{0, 1, \infty\}$$

is called the modular group. Now let $\widetilde{f}$ be the lift of f, so that $\widetilde{f} : \mathbb{D} \to \mathbb{D}$ and

$$f \circ \pi = \pi \circ \widetilde{f}.$$

For $z \in S^2 \setminus \{0, 1, \infty\}$, $\pi(\widetilde{z}) = z$. So the aim is to estimate $d_h(\widetilde{z}, \widetilde{f}(\widetilde{z}))$, which is at least $d(z, f(z))$, where $d(.,.)$ denoted the hyperbolic metric on $S^2 \setminus \{0, 1, \infty\}$.

The lift of f, $\widetilde{f}$ is the identity map on $\partial\mathbb{D}$ because f is homotopically trivial and fixes 0, 1 and ∞. From Theorem 3.4.3, we have

$$d_h(\widetilde{z}, \widetilde{f}(\widetilde{z})) \leq D(K),$$

and therefore

$$d(z, f(z)) \leq D(K).$$

We can use Theorem 3.4.3 because $\widetilde{f}$ is a homeomorphism (since f is), and is also K-quasiconformal. This proves the following theorem.

Theorem 3.4.4. *Let $\mathcal{F}$ be the family of quasiconformal maps of S^2 fixing* 0, 1 *and ∞. Then $\mathcal{F}$ is a compact subset of the topological space of all homeomorphisms of S^2 with respect to the C^0-topology.*

Exercise 3.4.3. Let $f : S^2 \to S^2$ be a homeomorphism and

$$\mathcal{F} = \{g = A \circ f \circ B : A, B \in \text{Möb}(S^2), g \text{ fixes } 0, 1, \infty\}.$$

Prove f is quasiconformal if and only if $\mathcal{F}$ is compact in the sense given above.

4

Quasi-isometries and quasisymmetric maps

Quasiconformal maps of the upper half-plane $\mathbb{H}$ to itself extend to the boundary $\mathbb{R}$. We show what conditions such mappings $\mathbb{R} \to \mathbb{R}$ must satisfy, and conversely that every such mapping extends to a quasiconformal mapping of $\mathbb{H}$. Such self-mappings of $\mathbb{R}$ are called quasisymmetric. We investigate properties of quasisymmetric maps and illustrate two ways of extending quasisymmetric maps to quasiconformal maps, the Beurling–Ahlfors extension and the conformally natural Douady–Earle extension. We also define the notion of a quasi-isometry and show that all quasiconformal maps are quasi-isometric.

4.1 Cross-ratio

Given a quadruplet of points z_1, z_2, z_3, z_4 in S^2, their cross-ratio is given by

$$\rho(z_1, z_2, z_3, z_4) = \frac{(z_3 - z_2)(z_1 - z_4)}{(z_2 - z_1)(z_4 - z_3)}.$$

If A is any Möbius transformation, then the cross-ratio has the property that for all quadruplets,

$$\rho(z_1, z_2, z_3, z_4) = \rho(A(z_1), A(z_2), A(z_3), A(z_4)),$$

and so if z_1, z_2, z_3, z_4 is any given quadruplet, we can apply a suitable Möbius transformation A to map three points to $0, 1, \infty$, and then

$$\rho(z_1, z_2, z_3, z_4) = \rho(0, 1, \infty, A(z_4)) = A(z_4).$$

Proposition 4.1.1. *Let f be a K-quasiconformal mapping. Given a non-empty compact set*

$$E \subset S^2 \setminus \{0, 1, \infty\},$$

then there exists a compact set

$$F \subset S^2 \setminus \{0, 1, \infty\}$$

such that for all quadruplets of points z_1, z_2, z_3, z_4 *which have cross-ratio*

$$\rho(z_1, z_2, z_3, z_4) \in E,$$

then

$$\rho(f(z_1), f(z_2), f(z_3), f(z_4)) \in F.$$

Proof. If this proposition is false, then there exists a sequence of quadruplets

$$\{(z_1^n, z_2^n, z_3^n, z_4^n)\}_{n=1}^{\infty}$$

such that $\rho(z_1^n, z_2^n, z_3^n, z_4^n) \in E$ for all n, but

$$\rho(f(z_1^n), f(z_2^n), f(z_3^n), f(z_4^n)) \to x \in \{0, 1, \infty\}.$$

Let A_n be the Möbius transformation that maps z_1^n, z_2^n, z_3^n to $0, 1, \infty$ respectively. Then $\rho(z_1^n, z_2^n, z_3^n, z_4^n) = A_n(z_4^n)$ is a sequence of points contained in E which, by passing to a subsequence if necessary, converges to $z \in E$.

Let B_n be the Möbius transformation which maps $f(z_1^n), f(z_2^n), f(z_3^n)$ to $0, 1, \infty$ respectively. Then $\rho(f(z_1^n), f(z_2^n), f(z_3^n), f(z_4^n)) = B_n(f(z_4^n))$ is a sequence which converges to $x \in \{0, 1, \infty\}$.

If $g_n = B_n \circ f \circ A_n^{-1}$, then g_n is a sequence of K-quasiconformal maps which fixes $0, 1, \infty$. By Theorem 3.4.4, and by passing if necessary to a subsequence, $g_n \to g$ uniformly on compact subsets of $S^2 \setminus \{0, 1, \infty\}$, and where g is a K-quasiconformal map. Since $g_n \to g$ uniformly on E,

$$\lim_{n \to \infty} g_n(A_n(z_4^n)) = g(z),$$

however $g_n(A_n(z_4^n)) \to x \in \{0, 1, \infty\}$, but since $g(x) = x$ and $z \neq x$, this contradicts g being a K-quasiconformal homeomorphism. □

4.2 Quasisymmetric maps

If $f : \mathbb{H} \to \mathbb{H}$ is a surjective quasiconformal map, then by Exercise 3.3.1, f can be extended to a homeomorphism of $\mathbb{R}$. Alternatively, f can be extended to the lower half-plane $\mathbb{H}^-$ by

$$f(z) = \overline{f(\overline{z})}$$

for $z \in \mathbb{H}^-$. This extension is a quasiconformal map on $\mathbb{C} \setminus \mathbb{R}$, but by Theorem 3.1.3, $\mathbb{R}$ is a removable Jordan curve, and so f can be extended to a quasiconformal map $\widetilde{f}$ on the whole of S^2. In particular, f extends to a quasiconformal map on the closure of $\mathbb{H}$, and $\widetilde{f}$ fixes ∞. For $X \in \mathbb{R}$ and $t \in \mathbb{R}^+$, consider the quadruple of points $X - t, X, X + t$, and ∞. By Proposition 4.1.1, there exists a constant $M > 0$ such that

$$\frac{1}{M} \leq \frac{\widetilde{f}(X+t) - \widetilde{f}(X)}{\widetilde{f}(X) - \widetilde{f}(X-t)} \leq M. \tag{4.1}$$

Definition 4.2.1. *A homeomorphism $\widetilde{f} : \mathbb{R} \to \mathbb{R}$ is M-quasisymmetric if equation (4.1) is satisfied for all $x \in \mathbb{R}$ and $t > 0$, and if $\widetilde{f}(\infty) = \infty$.*

Exercise 4.2.1. Suppose $\widetilde{f_n} : \mathbb{R} \to \mathbb{R}$ is a sequence of M-quasisymmetric functions, for a fixed $M > 0$, which fix 0 and 1. Prove that they form a normal family.

There is an analogous definition for quasisymmetric homeomorphisms of the circle. The following theorem shows that equation (4.1) turns out to be a necessary and sufficient condition for $\widetilde{f}$ to be an extension of a quasiconformal mapping.

Theorem 4.2.1. *A homeomorphism $\widetilde{f} : \mathbb{R} \to \mathbb{R}$ with $\widetilde{f}(\infty) = \infty$ is quasisymmetric if and only if it is the extension to $\mathbb{R}$ of some quasiconformal homeomorphism $f : \mathbb{H} \to \mathbb{H}$. Moreover, f can be chosen so that the maximal dilatation of f depends only on the quasisymmetry constant of $\widetilde{f}$.*

Proof. We have already seen above that every quasiconformal homeomorphism $f : \mathbb{H} \to \mathbb{H}$ extends to a quasisymmetric homeomorphism of the boundary. For the converse, which will only be sketched, we need to consider the Beurling–Ahlfors extension (see [7]) of a quasisymmetric homeomorphism $\widetilde{f}$ of $\mathbb{R}$ to all of $\overline{\mathbb{H}}$, given by

$$f(x+iy) = \frac{1}{2}\int_0^1 (\widetilde{f}(x+ty) + \widetilde{f}(x-ty))\,\mathrm{d}t + i\int_0^1 (\widetilde{f}(x+ty) - \widetilde{f}(x-ty))\,\mathrm{d}t. \tag{4.2}$$

On $\mathbb{R}$, we have that $f = \widetilde{f}$. Consider the functions

$$\alpha(x,y) = \int_0^1 \widetilde{f}(x+ty)\,\mathrm{d}t,$$

$$\beta(x,y) = \int_0^1 \widetilde{f}(x-ty)\,\mathrm{d}t.$$

Both α and β are continuously differentiable on $\mathbb{H}$, and the Jacobian of f, given by $\alpha_y\beta_x - \alpha_x\beta_y$ is positive on $\mathbb{H}$. This implies that f is a continuously differentiable bijective self-mapping of $\mathbb{H}$.

For any Möbius transformations A and B which fix $\mathbb{R}$, $B \circ f \circ A$ is the Beurling–Ahlfors extension of $B \circ \widetilde{f} \circ A|_{\mathbb{R}}$, and this composition with Möbius transformations does not change the maximal dilatation of f or the quasisymmetry constant of $\widetilde{f}$.

Suppose that there is not a constant $K(M)$ bounding the maximal dilatation of f. Then there exist M-quasisymmetric maps $\widetilde{f_n}$ and $z_n \in \mathbb{H}$ such that the dilatation quotients D_n of the Beurling–Ahlfors extensions f_n of $\widetilde{f_n}$ have the property that $D_n(z_n) \to \infty$. By composing with suitable Möbius transformations, we can assume that $\widetilde{f_n}$ fixes $0, 1, \infty$ and that $z_n = i$ for all n.

The $\widetilde{f_n}$ forms a normal family, and so $\widetilde{f_n}$ converges to some M-quasisymmetric function $\widetilde{f}$. Let f be the Beurling–Ahlfors extension of $\widetilde{f}$. If α_n and β_n are the corresponding functions defined above for $\widetilde{f_n}$, then we have

$$(\alpha_n)_x(i) = 1, \qquad (\beta_n)_x(i) = -\widetilde{f_n}(-1),$$

$$(\alpha_n)_y(i) = 1 - \int_0^1 \widetilde{f_n}(t)\,dt$$

and

$$(\beta_n)_y(i) = \widetilde{f_n}(-1) - \int_{-1}^0 \widetilde{f_n}(t)\,dt.$$

So, the partial derivatives of α_n and β_n converge to the partial derivatives of α and β at i, and in particular, the dilatation quotient $D_n(i)$ converges to the dilatation quotient of f at i, which contradicts the supposition that $D_n(i) \to \infty$. □

Remark. The maximal dilatation $K(M)$ of the Beurling–Ahlfors extension of a M-quasisymmetric map satisfies the condition that $K(M) \to 1$ as $M \to 1$. This can be seen from that fact that if $\widetilde{f_n}$ are quasisymmetric map with quasisymmetry constants $M_n \to 1$, then $\widetilde{f_n}$ converges to the identity mapping uniformly on compact subsets of $\mathbb{R}$. Since the Beurling–Ahlfors extension of the identity mapping on $\mathbb{R}$ is the identity on $\mathbb{H}$, the proof given above can be adapted to show that the maximal dilatations of the Beurling–Ahlfors extensions of $\widetilde{f_n}$ converge to 1.

Exercise 4.2.2. If $f : \partial\mathbb{D} \to \partial\mathbb{D}$, both f and f^{-1} are both of class C^1, then show that f is quasisymmetric.

Quasisymmetric maps can be viewed as one-dimensional versions of quasiconformal maps, but the following proposition, first proved by Carleson in [10], shows that they do not share all the same characteristics.

Proposition 4.2.1. *Quasisymmetric maps are not necessarily absolutely continuous.*

Proof. Let $f : M \to N$ be a quasiconformal map between closed hyperbolic Riemann surfaces which is not homotopic to a conformal map (we see in Chapter 7 that such things exist) and let $M \simeq \mathbb{H}/\Gamma$ and $N \simeq \mathbb{H}/\widetilde{\Gamma}$. Then f lifts to a quasiconformal map $\widetilde{f} : \mathbb{H} \to \mathbb{H}$ which respects the group structures, that is, for all $z \in \mathbb{H}, A \in \Gamma, B \in \widetilde{\Gamma}$,

$$(B \circ \widetilde{f} \circ A)(z) = \widetilde{f}(z).$$

Let $h : \mathbb{R} \to \mathbb{R}$ be the quasisymmetric extension of $\widetilde{f}$ to $\mathbb{R} = \partial\mathbb{H}$ and assume for a contradiction that h is absolutely continuous, and therefore has a derivative almost everywhere. Without loss of generality we can assume that $h(0) = 0$ and $h'(0) \neq 0$.

Let A_n be the Möbius transformation z/n, and B_n a Möbius transformation chosen so that

$$(B_n \circ h \circ A_n) : \mathbb{R} \to \mathbb{R}$$

fixes $0, 1, \infty$. Taking the limit as $n \to \infty$ is called rescaling h at the point $z = 0$. Since h is differentiable at 0 with non-zero derivative, this rescaling procedure gives that

$$\widehat{h} = \lim_{n\to\infty} (B_n \circ h \circ A_n)$$

is a linear map. To finish the proof, we need to show that $\widehat{h}$ is a quasisymmetric map which is an extension of a quasiconformal map g between Riemann surfaces which have covering groups given by, after passing to a subsequence n_i which for brevity we still denote as n,

$$\lim_{n\to\infty} (A_n^{-1} \circ \Gamma \circ A_n)$$

and

$$\lim_{n\to\infty} (B_n^{-1} \circ \widetilde{\Gamma} \circ B_n).$$

Then we will have the desired contradiction, since f was assumed not to be homotopic to a conformal map, whereas since $\widehat{h}$ is linear, g must be conformal, and g is homotopic to f.

To show that

$$\Gamma_n = (A_n^{-1} \circ \Gamma \circ A_n)$$

converges to the covering group of a closed Riemann surface, first observe that Γ is a Fuchsian group of the first kind, and is finitely generated, with say N generators, since M is closed. Since Γ and $\widetilde{\Gamma}$ are covering groups of quasiconformally related Riemann surfaces, they have the same number of generators. The orbit of $i \in \mathbb{H}$ under Γ is well distributed, in the sense that there is a constant C such that every disk of hyperbolic radius C in $\mathbb{H}$ contains at least one point of the orbit of i under Γ.

Let D_i be a Dirichlet fundamental region for Γ constructed from $i \in \mathbb{H}$ (see [31] for the definition of a Dirichlet region). For any $w \in \mathbb{H}$, there is a unique $z \in D_i$ such that the Dirichlet fundamental region D_w constructed from w is a translate of D_z under some $\gamma \in \Gamma$. For each n, consider the Dirichlet fundamental region for Γ_n constructed from the $z_n \in D_i$ which corresponds to $\Gamma_n(i)$. Since D_i is relatively compact in $\mathbb{H}$, by passing to a subsequence, z_n converges and hence D_{z_n} converges.

Since A_n is a Möbius transformation, each fundamental region D_{z_n} for Γ_n is a hyperbolic polygon with N sides, and is generated by Möbius transformations $g_1^n, \dots, g_N^n$ which permute the sides of the polygon D_{z_n}. Since the fundamental regions D_{z_n} converge, the generators of Γ_n, the g_i^n, must also converge to some group Γ_∞ generated by $g_1^\infty, \dots, g_N^\infty$. This means that Γ_∞ is a finitely generated Fuchsian group of the first kind, and hence the Riemann surface given by $\mathbb{H}/\Gamma_\infty$ is a closed surface. Similarly $(B_n^{-1} \circ \widetilde{\Gamma} \circ B_n)$ converges and the proof is complete. □

Remark. The ideas contained in this proof can be used to prove the following Mostow rigidity theorem [38]. For an introduction to hyperbolic geometry see, for example, [4].

Theorem 4.2.2. *Let M and M' be closed hyperbolic manifolds of dimension ≥ 3, and let $F : M \to M'$ be a homeomorphism. Then F is homotopic to an isometry.*

We sketch how the proof runs. For the case $n = 3$, since both $M \simeq \mathbb{B}^3/\Gamma$ and $M' \simeq \mathbb{B}^3/\Gamma'$ are hyperbolic, the map F can be lifted to a map $\widetilde{F} : \mathbb{B}^3 \to \mathbb{B}^3$, where $\mathbb{B}^3$ is the unit ball of dimension 3, and such that $\widetilde{F}$ extends to a quasiconformal map $f : S^2 \to S^2$. Then since f is differentiable almost everywhere, without loss of generality $f'(0) \neq 0$. Rescaling as in Proposition 4.2.1 using suitable Möbius transformations A_n and B_n, we conclude that $f_n = B_n \circ f \circ A_n^{-1}$ converges to a smooth map between hyperbolic 3-manifolds

$$\lim_{n\to\infty} \frac{\mathbb{B}^3}{(A_n^{-1} \circ \Gamma \circ A_n)}$$

and

$$\lim_{n\to\infty} \frac{\mathbb{B}^3}{(B_n^{-1} \circ \Gamma' \circ B_n)}.$$

The corresponding maps $\widetilde{F_n}$ converge to a limiting homeomorphism $\widetilde{F}$, homotopic to F, which must be an isometry since f_n converges to a linear Möbius transformation.

4.3 Quasi-isometry

Definition 4.3.1. *Let $f : \mathbb{H} \to \mathbb{H}$. We say that f is (K,a)-quasi-isometric if there exist constants K and $a > 0$ such that*

$$\frac{d_h(z,w)}{K} - a \leq d_h(f(z), f(w)) \leq K d_h(z,w) + a.$$

for all $z, w \in \mathbb{H}$ and where d_h is the hyperbolic distance on $\mathbb{H}$.

The following Lemma 4.3.1, proved in [33], shows that quasi-isometries map every hyperbolic geodesic to a neighbourhood of another hyperbolic geodesic.

Lemma 4.3.1. *Let $f : \mathbb{H} \to \mathbb{H}$ be (K,a)-quasi-isometric and let $\gamma \subset \mathbb{H}$ be a hyperbolic geodesic. Then there is another geodesic $\widetilde{\gamma}$ and a constant $C(K,a)$ such that $f(\gamma)$ is in the $C(K,a)$-neighbourhood of $\widetilde{\gamma}$.*

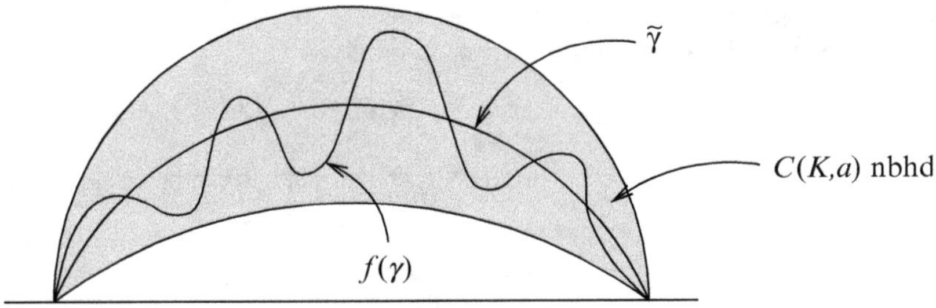

Proof. Let γ be parameterized by $t \in \mathbb{R}$ and assume that $f \circ \gamma$ is continuous. For large $n > 0$, let γ_n be the complete geodesic which passes through $f \circ \gamma(-n)$ and $f \circ \gamma(n)$. Now consider the neighbourhood $B_r(\gamma_n)$ of γ_n of radius r. Let the interval (x, y) be a maximal interval for which $f \circ \gamma(x, y)$ is not contained in $B_r(\gamma_n)$. This means that $f \circ \gamma(x)$ and $f \circ \gamma(y)$ lie on the boundary of $B_r(\gamma_n)$, and by a standard fact of hyperbolic geometry, the nearest point projection of $f \circ \gamma(x, y)$ to γ_n is contracting by a factor of $\cosh(r)$. Thus

$$\frac{|x-y|}{K} \le d_h(f \circ \gamma(x), f \circ \gamma(y)) \le 2r + \frac{d_h(f \circ \gamma(x), f \circ \gamma(y))}{\cosh(r)} \le 2r + \frac{K|x-y|}{\cosh(r)}.$$

If r is taken large enough that

$$\cosh(r) \gg K^2,$$

then the inequality above implies that $|x - y|$ cannot get too large and hence $f \circ \gamma(x, y)$ stays close to γ_n. In particular, there is an absolute constant θ such that

$$f \circ \gamma(-n, n) \subseteq B_\theta(\gamma_n).$$

Since the space of geodesics within distance θ of $f \circ \gamma(0)$ is compact, then by passing to a subsequence of $\{\gamma_n\}_{n \in \mathbb{N}}$ yields a geodesic $\widetilde{\gamma}$ which is within distance θ of $f \circ \gamma$. Since distinct geodesics diverge in the hyperbolic metric, $\widetilde{\gamma}$ is unique. □

Theorem 4.3.1. *Let $f : \mathbb{H} \to \mathbb{H}$ be a (K, a)-quasi-isometry normalized to fix ∞. Then f extends to continuously to $h : \overline{\mathbb{R}} \to \overline{\mathbb{R}}$ and $f|_{\overline{\mathbb{R}}} = h$. The map h is quasisymmetric.*

Proof. Let γ be a hyperbolic geodesic with endpoint at $x_0 \in \mathbb{R}$. Then $f(\gamma)$ is contained in a $C(K, a)$-neighbourhood of another geodesic $\widetilde{\gamma}$. Define $h(x_0)$ to be the endpoint of $\widetilde{\gamma}$. To remove the ambiguity of which endpoint, consider another geodesic γ_1 with endpoint at x_0 and its image, which is contained in a neighbourhood of $\widetilde{\gamma_1}$, and use the endpoint common to both $\widetilde{\gamma}$ and $\widetilde{\gamma_1}$. Define h on all of $\mathbb{R}$ in a similar fashion, and also define $h(\infty) = \infty$.

Now, for a contradiction, assume h is not quasisymmetric. This means that there exist sequences $(x_n) \in \mathbb{R}, (t_n) \in \mathbb{R}^+$ such that

$$\left| \frac{h(x_n + t_n) - h(x_n)}{h(x_n) - h(x_n - t_n)} \right| \to \infty.$$

The expression given above could degenerate to 0, but this case is treated analogously. Now define the Möbius transformations

$$T_n(x) = x + x_n, \qquad \widetilde{T_n}(x) = x - h(x_n),$$

$$A_n(x) = t_n x, \qquad B_n(x) = \frac{x}{|h(x_n + t_n) - h(x_n)|}$$

and consider

$$f_n(x) = (B_n \circ \widetilde{T_n} \circ f \circ T_n \circ A_n)(x)$$

and

$$h_n(x) = (B_n \circ \widetilde{T_n} \circ h \circ T_n \circ A_n)(x).$$

We have that f_n is (K, a)-quasi-isometric because f is and the other maps in the composition are isometries. Also, h_n fixes $0, 1$ and ∞. By the assumption,

$$\left| \frac{h(x_n + t_n) - h(x_n)}{h(x_n) - h(x_n - t_n)} \right| = \left| \frac{1}{h_n(-1)} \right| \to \infty.$$

Let γ_1 be the geodesic joining 1 and ∞. As in Lemma 4.3.1, consider $\widetilde{\gamma_1}$, which is the geodesic within a bounded distance of the quasi-geodesic $f_n(\gamma_1)$. Since h_n fixes 1 and ∞, $\widetilde{\gamma_1}$ is just γ_1. Let γ_2 be the geodesic joining -1 and 0. Similarly, $\widetilde{\gamma_2}$ is the geodesic joining $h_n(-1)$ and 0. Now for two geodesics γ and γ', define their distance as

$$d_h(\gamma, \gamma') = \inf_{z \in \gamma, w \in \gamma'} d_h(z, w).$$

In particular, $d_h(\gamma_1, \gamma_2)$ is finite. By Lemma 4.3.1 and since f_n is (K, a)-quasi-isometric,

$$d_h(\widetilde{\gamma_1}, \widetilde{\gamma_2}) - 2C(K) \leq d_h(f_n(\gamma_1), f_n(\gamma_2)) \leq K d_h(\gamma_1, \gamma_2) + a$$

which implies

$$d_h(\widetilde{\gamma_1}, \widetilde{\gamma_2}) \leq K d_h(\gamma_1, \gamma_2) + 2C(K) + a.$$

The left-hand side of this equation is just a fixed constant. But

$$d_h(\widetilde{\gamma_1}, \widetilde{\gamma_2}) = d_h(\gamma_1, \widetilde{\gamma_2})$$

and $\widetilde{l_2}$ depends on n via $h_n(-1)$. Now since as $n \to \infty$, $h_n(-1) \to 0$, then $d_h(\widetilde{\gamma_1}, \widetilde{\gamma_2}) \to \infty$, which is a contradiction. □

Proposition 4.3.1. *Let $f, g : \mathbb{H} \to \mathbb{H}$ be two (K, a)-quasi-isometries which agree on the boundary. Then there is a constant $D(K, a)$ such that*

$$d_h(f(z)), g(z)) \leq D(K, a)$$

for all $z \in \mathbb{H}$.

Proof. Choose geodesics l_1, l_2 which intersect at $z \in \mathbb{H}$. Then $f(l_i), g(l_i)$ are contained in a $C(K)$-neighbourhood of $\widetilde{l_i}$ for $i = 1, 2$ (since f and g agree on the boundary, we get the same geodesics $\widetilde{l_i}$ for $f(l_i)$ and $g(l_i)$). Now let $\widetilde{l_1}$ and $\widetilde{l_2}$ intersect at w. Then

$$d_h(f(z), w) \leq 2C(K, a),$$

and

$$d_h(g(z), w) \leq 2C(K, a),$$

and so

$$d_h(f(z), g(z)) \leq 4C(K, a) = D(K, a).$$

□

The following theorem shows that quasiconformal mappings are quasi-isometric. This was first proved in [40], and improved in [18] to show that if f is K-quasiconformal, then the same constant K holds to satisfy the quasi-isometry condition.

Theorem 4.3.2. *Let $f : \mathbb{D} \to \mathbb{D}$ be K-quasiconformal. Then there exists a constant $a(K) > 0$ depending only on K such that*

$$\frac{d(z, w)}{K} - a(K) \leq d(f(z), f(w)) \leq Kd(z, w) + a(K).$$

That is, f is quasi-isometric.

Proof. Without loss of generality, assume that f fixes 0. By Theorem 3.2.1,

$$|f(z)| \leq \varphi_K(|z|)$$

where φ_K is the extremal domain function considered in Section 3.2. It can be shown that for $K = 2$,

$$\varphi_2(t) = \frac{2\sqrt{t}}{1 + t}$$

for $0 \leq t \leq 1$, and so for $K = 2$,

$$|f(z)| \leq \frac{2\sqrt{|z|}}{1 + |z|}.$$

Recall that the hyperbolic distance between 0 and $z \in \mathbb{D}$ is given by the formula

$$d_h(0, z) = \log \frac{1 + |z|}{1 - |z|},$$

and also consider the strictly increasing function $g : [0, 1] \to [1, 2]$ given by

$$g(t) = \frac{(1 + t)^2}{1 + t^2}.$$

Then

$$d_h(0, f(z)) - 2d_h(0, z) \leq 2 \log g(\sqrt{|z|}) < 2 \log 2$$

which implies that

$$d_h(0, f(z)) \leq 2d_h(0, z) + 2\log 2.$$

This proves one inequality in the statement of the theorem for $K = 2$. The other inequality is obtained by considering f^{-1}. To extend the result for other values of K, we need to use the following fact. Namely, that given $L \geq 1$ and a point p on a hyperbolic geodesic γ, there exists an L-quasiconformal diffeomorphism $f_L : \mathbb{D} \to \mathbb{D}$ fixing p, mapping γ to itself and stretching hyperbolic distances along γ by a factor of L, and stretching all other distances by a factor between 1 and L.

For $1 < K < 2$, apply this fact with $L = 2/K$, $p = 0$ and γ being the geodesic passing through 0 and $f(z)$. Then $f_L f$ is 2-quasiconformal and fixes 0. Therefore,

$$Ld_h(0, f(z)) = d_h(0, f_L f(z)) \leq 2d_h(0, z) + 2\log 2$$

$$d_h(0, f(z)) \leq Kd_h(0, z) + K\log 2,$$

which finishes off the case $1 < K < 2$. For $K > 2$, find $n \geq 2$ such that $2^{n-1} < K \leq 2^n$. Then f can be written as a composition of $K^{1/n}$-quasiconformal maps $f_i : \mathbb{D} \to \mathbb{D}$, $f = f_1 \circ \cdots \circ f_n$. Apply the inequality to each f_i with K replaced by $K^{1/n} \leq 2$. There then exists a constant a_n such that for all $z, w \in \mathbb{D}$,

$$d_h(f(z), f(w)) \leq Kd_h(z, w) + a_n.$$

An explicit calculation shows that

$$a_n = \frac{K^{1/n}(K-1)\log 2}{K^{1/n}-1} \leq \frac{(K-1)\sqrt{2}\log 2}{\sqrt{2}-1}.$$

This completes the proof. □

Theorems 4.3.1 and 4.3.2 together also give a different proof that a quasiconformal map from $\mathbb{D}$ to $\mathbb{D}$ extends continuously to $\partial\mathbb{D}$, and the extension is a quasisymmetric map.

Theorem 4.3.3. *Let M and N be closed hyperbolic Riemann surfaces and let $f : M \to N$ be K-quasiconformal. If γ is a geodesic on M, and $\widetilde{\gamma}$ is the geodesic on N which is homotopic to $f(\gamma)$. Then*

$$\frac{1}{K} \leq \frac{|\widetilde{\gamma}|}{|\gamma|} \leq K,$$

where $|\gamma|$ denotes the length of γ.

This was first proved in [56].

Proof. Let $\pi_M : \mathbb{D} \to M$ and $\pi_N : \mathbb{D} \to N$ be the projections from the universal cover. Let $\widetilde{f} : \mathbb{D} \to \mathbb{D}$ be a lift of f, and choose a component γ^* of $\pi_M^{-1}(\gamma)$. Now γ^* can be written as a disjoint union of half-closed arcs

$$\gamma^* = \cup_j \gamma_j$$

where $\pi_M : \gamma_j \to \gamma$ is bijective. Apply Theorem 4.3.2 to give

$$|\widetilde{f}(\gamma_j \cup \gamma_{j+1})| \leq K|\gamma_j \cup \gamma_{j+1}| + a(K).$$

Projecting back to the surfaces,

$$2|\widetilde{\gamma}| \leq 2K|\gamma| + a(K)$$

or

$$|\widetilde{\gamma}| \leq K|\gamma| + \frac{a(K)}{2}.$$

So in general, for any n,

$$|\widetilde{\gamma}| \leq K|\gamma| + \frac{a(K)}{n}.$$

Let $n \to \infty$ and we can conclude that

$$|\widetilde{\gamma}| \leq K|\gamma|.$$

□

4.4 The barycentric extension

The main theorem of this section shows that quasisymmetric homeomorphisms from $\mathbb{T}$ to $\mathbb{T}$ can be extended to quasiconformal self-mappings of $\mathbb{D}$ in a conformally natural way. This was proved by Douady and Earle in [12], and our presentation of their result follows [42]. Further results in this area can be found in [1] and [36]. Recall the Beurling–Ahlfors extension in (4.2), which extends a quasisymmetric homeomorphism $\widetilde{f}$ of $\mathbb{R}$ to a quasiconformal mapping f of $\mathbb{H}$,

$$f(x+iy) = \frac{1}{2}\int_0^1 (\widetilde{f}(x+ty) + \widetilde{f}(x-ty))\,\mathrm{d}t + i\int_0^1 (\widetilde{f}(x+ty) - \widetilde{f}(x-ty)\,\mathrm{d}t.$$

This was the first extension of a quasisymmetric mapping to a quasiconformal one; however, it is not conformally natural in the way described below.

Theorem 4.4.1. *A sense-preserving homeomorphism $\varphi : \mathbb{T} \to \mathbb{T}$ can be extended to a homeomorphism $\widetilde{\varphi} : \overline{\mathbb{D}} \to \overline{\mathbb{D}}$ that is real analytic in $\mathbb{D}$ and such that*

(a) *if σ, τ are Möbius transformations of $\mathbb{D}$, then the extension of $\sigma \circ \varphi \circ \tau$ is given by $\sigma \circ \widetilde{\varphi} \circ \tau$,*

(b) *if φ is quasisymmetric, then $\widetilde{\varphi}$ is quasiconformal in $\mathbb{D}$.*

This extension is called the barycentric or Douady–Earle extension. The function $\widetilde{\varphi}$ is real analytic if its real and imaginary parts are real analytic functions. The barycentric extension is uniquely determined by the conformal naturalness condition (a) above, together with the condition that

$$\int_{\mathbb{T}} \varphi(\zeta)|\mathrm{d}\zeta| = 0 \implies \widetilde{\varphi}(0) = 0.$$

We first give some consequences of Theorem 4.4.1 before proving it at the end of the section.

Theorem 4.4.2. *Every quasisymmetric map h of $\mathbb{T}$ into $\mathbb{C}$ can be extended to a quasiconformal map $\widetilde{h}$ of $\widehat{\mathbb{C}}$ onto $\widehat{\mathbb{C}}$ such that the extension of $\sigma \circ h \circ \tau$, for a Möbius transformation σ and a Möbius transformation τ of $\mathbb{D}$ onto $\mathbb{D}$, is given by $\sigma \circ \widetilde{h} \circ \tau$.*

Proof. Since h is quasisymmetric, $h(\mathbb{T})$ is a quasicircle, which is assumed to be positively oriented. Let f and g map $\mathbb{D}$ and $\mathbb{D}^* = \mathbb{C} \setminus \overline{\mathbb{D}}$ onto the inner and outer domains of $h(\mathbb{T})$ such that g fixes ∞. The restrictions of f and g to $\mathbb{T}$ are quasisymmetric and hence $\varphi = f^{-1} \circ h$ and $\psi = g^{-1} \circ h$ are quasisymmetric.

Let $\widetilde{\varphi}$ be the barycentric extension of φ to $\overline{\mathbb{D}}$, and $\widetilde{\psi}$ be the barycentric extension of ψ to $\overline{\mathbb{D}^*}$. Define

$$\begin{aligned} \widetilde{h}(z) &= f \circ \widetilde{\varphi}(z), \qquad |z| \leq 1, \\ \widetilde{h}(z) &= g \circ \widetilde{\psi}(z), \qquad |z| \geq 1. \end{aligned} \tag{4.3}$$

On $\mathbb{T}$, these definition agree, and $\widetilde{h}(z) = h(z)$ there. Thus $\widetilde{h}$ is a homeomorphism of $\widehat{\mathbb{C}}$ onto $\widehat{\mathbb{C}}$, and this function is real analytic in $\mathbb{C} \setminus \mathbb{T}$. If $|z| < 1$, then

$$\widetilde{h}_{\overline{z}} = f'(\widetilde{\varphi})\widetilde{\varphi}_{\overline{z}}$$

and

$$\widetilde{h}_z = f'(\widetilde{\varphi})\widetilde{\varphi}_z.$$

Since

$$\left|\frac{\widetilde{\varphi}_{\overline{z}}}{\widetilde{\varphi}_z}\right| \leq K < 1$$

from Theorem 4.4.1 (b), then

$$\left|\frac{\widetilde{h}_{\overline{z}}}{\widetilde{h}_z}\right| \leq K < 1.$$

The case for $|z| > 1$ follows analogously. Let σ be a Möbius transformation and τ a Möbius transformation of $\mathbb{D}$ to $\mathbb{D}$. Therefore τ is also a Möbius transformation of $\mathbb{D}^*$ onto $\mathbb{D}^*$. Consider $h^* = \sigma \circ h \circ \tau$ and the conformal maps $f^* = \sigma \circ f \circ \tau$ and $g^* = \sigma \circ g \circ \tau$. Then, analogously to the first part of the proof, $\varphi^* = \tau^{-1} \circ \varphi \circ \tau$ and $\psi^* = \tau^{-1} \circ \psi \circ \tau$. Hence from the conformal naturalness condition, for $|z| \leq 1$ we have

$$\widetilde{h^*}(z) = f^* \circ \widetilde{\varphi^*}(z) = \sigma \circ f \circ \widetilde{\varphi} \circ \tau(z)$$

and for $|z| \geq 1$,

$$\widetilde{h^*}(z) = g^* \circ \widetilde{\psi^*}(z) = \sigma \circ g \circ \widetilde{\psi} \circ \tau(z).$$

Hence $\widetilde{h^*} = \sigma \circ \widetilde{h} \circ \tau$. □

Theorem 4.4.3. *Every conformal map f of $\mathbb{D}$ onto a quasidisk can be extended to a quasiconformal map $\widetilde{f}$ of $\widehat{\mathbb{C}}$ onto $\widehat{\mathbb{C}}$. The extension of $\sigma \circ f \circ \tau$ where σ is a Möbius transformation and τ is a Möbius transformation of $\mathbb{D}$ onto $\mathbb{D}$ is $\sigma \circ \widehat{f} \circ \tau$.*

Proof. Choose h in the proof of Theorem 4.4.2 as the restriction of f to $\mathbb{T}$, which is quasisymmetric. Then φ in the proof of Theorem 4.4.2 is the identity, and so is $\widetilde{\varphi}$. Thus the quasiconformal extension of f to $\widehat{\mathbb{C}}$ is given by (4.3). □

We now move on giving the construction for the barycentric extension. Let $\varphi : \mathbb{T} \to \mathbb{T}$ be a sense-preserving homeomorphism. The barycentric extension is based on the function

$$F(z,w) \equiv F_\varphi(z,w) = \int_{\mathbb{T}} \frac{w - \varphi(\zeta)}{1 - \overline{w}\varphi(\zeta)} p(z,\zeta)|\mathrm{d}\zeta|, \tag{4.4}$$

for $z, w \in \mathbb{D}$ and where $p(z,\zeta)$ is the Poisson kernel

$$p(z,\zeta) = \frac{1}{2\pi}\frac{1-|z|^2}{|\zeta - z|^2}.$$

Since

$$\int_{\mathbb{T}} p(z,\zeta)|\mathrm{d}\zeta| = 1,$$

we can rewrite (4.4) as

$$F(z,w) = w - \frac{1-|w|^2}{2\pi} \int_{\mathbb{T}} \frac{\varphi(\zeta)}{1-\overline{w}\varphi(\zeta)} \frac{1-|z|^2}{|\zeta - z|^2} |\mathrm{d}\zeta|. \tag{4.5}$$

Now let σ and τ be Möbius transformations from $\mathbb{D}$ to $\mathbb{D}$. Then

$$\frac{\sigma(w) - \sigma(\xi)}{1 - \overline{\sigma(w)}\sigma(\xi)} q(w) = \frac{w - \xi}{1 - \overline{w}\xi}$$

where $|q(w)| = 1$ and q is independent of ξ. With the substitution $\zeta = \tau(s)$ in (4.4),

$$\begin{aligned} F(\tau(z),w) &= \int_{\mathbb{T}} \frac{w - \varphi(\tau(z))}{1 - \overline{w}\varphi(\tau(s))} p(\tau(z),\tau(s))|\tau'(s)||\mathrm{d}s| \\ &= q(w) \int_{\mathbb{T}} \frac{\sigma(w) - (\sigma \circ \varphi \circ \tau)(s)}{1 - \overline{\sigma(w)}(\sigma \circ \varphi \circ \tau)(s)} p(z,s)|\mathrm{d}s| \end{aligned}$$

and therefore

$$F(\tau(z), \sigma^{-1}(w)) = \widetilde{q}(w) F_{\sigma\circ\varphi\circ\tau}(z,w), \tag{4.6}$$

where $|\widetilde{q}(w)| = 1$. This allows information at $(0,0)$ to be transferred to any point of $\mathbb{D} \times \mathbb{D}$. We now consider the local behaviour of F at $(0,0)$. The Fourier coefficients of φ are

$$a_n = \frac{1}{2\pi} \int_{\mathbb{T}} \overline{\zeta^n} \varphi(\zeta) |d\zeta|$$

for $n \in \mathbb{Z}$. We assume that $a_0 = 0$. The partial derivatives satisfy

$$\begin{aligned} F_z(0,0) &= -\frac{1}{2\pi} \int_{\mathbb{T}} \overline{\zeta} \varphi(\zeta) |d\zeta| = -a_1, \\ F_{\overline{z}}(0,0) &= -\frac{1}{2\pi} \int_{\mathbb{T}} \zeta \varphi(\zeta) |d\zeta| = -a_{-1}, \\ F_w(0,0) &= 1, \end{aligned}$$

and

$$F_{\overline{w}}(0,0) = -\frac{1}{2\pi} \int_{\mathbb{T}} \varphi(\zeta)^2 |d\zeta| = -2 \sum_{n=1}^{\infty} a_n a_{-n} \equiv b.$$

It can be shown that (see [42])

$$\delta \equiv |a_1|^2 - |a_{-1}|^2 > 0,$$

and so, using Parseval's theorem and the fact that $|\varphi(\zeta)| = 1$, we have

$$\begin{aligned} |b|^2 \le |b| &\le 2|a_1 a_{-1}| + \sum_{n=2}^{\infty} (|a_n|^2 + |a_{-n}|^2) \\ &= 1 - (|a_1| - |a_{-1}|)^2 \le 1 - \frac{\delta^2}{4}. \end{aligned}$$

Therefore, the Jacobian satisfies

$$1 \ge J_w(0,0) = |F_w(0,0)|^2 - |F_{\overline{w}}(0,0)|^2 = 1 - |b|^2 \ge \frac{\delta^2}{4}.$$

Since $F(0,0) = 0$, because $a_0 = 0$, the implicit function theorem implies that $F(z,w) = 0$ has a unique solution $w = \widetilde{\varphi}(z)$ near 0 with $\widetilde{\varphi}(0) = 0$. The partial derivatives of $\widetilde{\varphi}$ are

$$\widetilde{\varphi}_z = \frac{\overline{(F_{\overline{z}})} F_{\overline{w}} - F_z \overline{(F_w)}}{J_w}$$

and

$$\widetilde{\varphi}_{\overline{z}} = \frac{\overline{(F_z)} F_{\overline{w}} - F_{\overline{z}} \overline{(F_w)}}{J_w},$$

and so

$$\widetilde{\varphi}_z(0) = \frac{a_1 - \overline{a_{-1}}b}{1-|b|^2},$$

$$\widetilde{\varphi}_{\overline{z}}(0) = \frac{a_{-1} - \overline{a_1}b}{1-|b|^2}.$$

Therefore,

$$|\widetilde{\varphi}_z(0)|^2 - |\widetilde{\varphi}_{\overline{z}}(0)|^2 = \frac{|a_1|^2 - |a_{-1}|^2}{1-|b|^2} \geq \delta$$

and so we can conclude that $\widetilde{\varphi}$ is a local diffeomorphism at 0.

Lemma 4.4.1. *Given $z \in \mathbb{D}$, there exists a unique $w \in \mathbb{D}$ such that $F_\varphi(z,w) = 0$.*

Proof. Fix $z \in \mathbb{D}$. From (4.5),

$$|F(z,w) - w| \leq \frac{1}{2\pi}\int_{\mathbb{T}} \frac{1-|w|^2}{|\varphi(\zeta) - w|}\frac{1-|z|^2}{|\zeta - z|^2}|\mathrm{d}\zeta|.$$

Now let $w_0 \in \mathbb{T}$ and $w_n \in \mathbb{D}$ such that $w_n \to w_0$. For $\zeta \notin \varphi^{-1}(w_0)$, the integrand conerges to 0, and since it is bounded by $4/(1-|z|)$, Lebesgue's bounded convergence theorem implies that as $n \to \infty$,

$$|F(z,w_n) - w_n| \to 0.$$

Therefore, $F(z,w)$ is continuous in $w \in \overline{\mathbb{D}}$ if we define $F(z,w) = w$ for $w \in \mathbb{T}$. Then the function $w \mapsto F(z,w)$ has positive Jacobian for all w with $F(z,w) = 0$. The result follows from the Poincaré–Hopf theorem, which we state here: if a vector field which is continuous in $\overline{\mathbb{D}}$ and continuously differentiable in $\mathbb{D}$ points outwards at all points of $\mathbb{T}$ with positive Jacobian at all its zeros, then it has exactly one zero in $\mathbb{D}$. □

We are now in a position to construct the barycentric extension of a sense-preserving homeomorphism $\varphi : \mathbb{T} \to \mathbb{T}$. Define

$$\widetilde{\varphi}(z) = \begin{cases} \varphi(z), & z \in \mathbb{T} \\ w, \text{ where } F_\varphi(z,w) = 0, & z \in \mathbb{D}. \end{cases}$$

This function $\widetilde{\varphi}$ is the barycentric extension of φ and is uniquely determined by the equation

$$F_\varphi(z, \widetilde{\varphi}(z)) = \frac{1}{2\pi}\int_T \frac{\widetilde{\varphi}(z) - \varphi(\zeta)}{1 - \overline{\widetilde{\varphi}(z)}\varphi(\zeta)}\frac{1-|z|^2}{|\zeta - z|^2}|\mathrm{d}\zeta| = 0.$$

Now, the complex harmonic extension of φ is given by

$$\varphi^*(z) = \int_{\mathbb{T}} \varphi(\zeta)p(z,\zeta)\,|\mathrm{d}\zeta|$$

for $z \in \mathbb{D}$, and where $\varphi^*(z) = \varphi(z)$ for $z \in \mathbb{T}$. The complex harmonic extension is continuous in $\overline{\mathbb{D}}$ and is a homeomorphism of $\overline{\mathbb{D}}$ onto $\overline{\mathbb{D}}$. Before proving Theorem 4.4.1, we need a lemma which we will state and not prove. The proof can be found in [42].

Lemma 4.4.2. *Write $\varphi(\mathrm{e}^{it}) = \mathrm{e}^{iu(t)}$ where u is real, continuous and 2π-periodic. Suppose*

$$u(t+s) - u(t) \geq \alpha \quad \textit{for} \quad \frac{\pi}{3} \leq s \leq \frac{2\pi}{3}, t \in \mathbb{R}, \quad \textit{with} \quad 0 < \alpha \leq \frac{\pi}{3}. \tag{4.7}$$

Then

$$\delta = |a_1|^2 - |a_{-1}|^2 \geq \frac{2}{\pi} \sin^3 \alpha > 0$$

and

$$\left| \frac{\widetilde{\varphi}_{\overline{z}}(0)}{\widetilde{\varphi}_z(0)} \right| \leq 1 - \frac{1}{125} \sin^9 \alpha.$$

We now prove Theorem 4.4.1.

Proof. (a) Firstly, we have to show the conformal naturalness property of the barycentric extension. For $\zeta_0 \in \mathbb{T}$,

$$\widetilde{\varphi}(z) - \varphi^*(z) = \int_{\mathbb{T}} \frac{\varphi(\zeta) - \widetilde{\varphi}(z)}{1 - \overline{\widetilde{\varphi}(z)}\varphi(\zeta)} \overline{\widetilde{\varphi}(z)} (\varphi(\zeta) - \varphi(\zeta_0)) p(z,\zeta)\, |\mathrm{d}\zeta|$$

and so

$$|\widetilde{\varphi}(z) - \varphi^*(z)| \leq \int_{\mathbb{T}} |\varphi(\zeta) - \varphi(\zeta_0)| p(z,\zeta)\, |\mathrm{d}\zeta|. \tag{4.8}$$

If we let $z \in \mathbb{D}$ vary and converge to ζ_0, then the right-hand side of (4.8) converges to 0 since φ is continuous. Since φ^* is continuous in $\overline{\mathbb{D}}$, $\widetilde{\varphi}$ must also be continuous in $\overline{\mathbb{D}}$. The conformal naturalness of $\widetilde{\varphi}$ is now a consequence of (4.6) and the definition of $F_\varphi(z, \widetilde{\varphi}(z))$.

Now given $z_0 \in \mathbb{D}$, choose σ and τ to be Möbius transformations from $\mathbb{D}$ to $\mathbb{D}$ such that $\tau(0) = z_0$ and $\sigma(0) = \widetilde{\varphi}(z_0)$, and define $h = \sigma^{-1} \circ \widetilde{\varphi} \circ \tau$. Then $h(0) = 0$ and

$$h_z(0) = \widetilde{\varphi}_z(z_0) \frac{\tau'(0)}{\sigma'(0)},$$

$$h_{\overline{z}}(0) = \widetilde{\varphi}_{\overline{z}}(z_0) \frac{\overline{\tau'(0)}}{\sigma'(0)}.$$

Thus

$$|\widetilde{\varphi}_z(z_0)|^2 - |\widetilde{\varphi}_{\overline{z}}(z_0)|^2 > 0.$$

Since $\widetilde{\varphi}(\mathbb{D}) \subseteq \mathbb{D}$ and $\widetilde{\varphi}(\mathbb{T}) = \mathbb{T}$, then we can conclude that $\widetilde{\varphi}$ is an unbranched proper covering and thus a homeomorphism of $\overline{\mathbb{D}}$ onto $\overline{\mathbb{D}}$.

(b) Now let us assume that φ is quasisymmetric. Then

$$\left|\frac{\widetilde{\varphi}_{\overline{z}}(z_0)}{\widetilde{\varphi}_z(z_0)}\right| = \left|\frac{h_{\overline{z}}(0)}{h_z(0)}\right|. \tag{4.9}$$

Write $h = \mathrm{e}^{iu}$ and suppose (4.7) does not hold in Lemma 4.4.2, so that

$$u(t+s) - u(t) < \alpha$$

for some t, s with $\pi/3 \leq s \leq 2\pi/3$. Then

$$|h(\mathrm{e}^{i(t+s)}) - h(\mathrm{e}^{it})| < 2\sin\left(\frac{\alpha}{2}\right) < \alpha.$$

If

$$z_1 = \mathrm{e}^{it+is}, \qquad z_2 = \mathrm{e}^{it+10i\pi/9},$$
$$z_3 = \mathrm{e}^{it+11i\pi/9}, \qquad z_4 = \mathrm{e}^{it},$$

then

$$\frac{|z_1 - z_3|\,|z_2 - z_4|}{|z_1 - z_4|\,|z_2 - z_3|} \leq \frac{2}{\sin(\pi/18)} = C_1,$$

where C_1 denotes a constant. Since $\widetilde{\varphi}$ is quasisymmetric, and $h = \sigma^{-1} \circ \widetilde{\varphi} \circ \tau$, then

$$|h(z_1) - h(z_3)||h(z_2) - h(z_4)| < C_2\alpha.$$

Therefore,

$$\min(|h(z_1) - h(z_3)|, |h(z_2) - h(z_4)|) < (C_2\alpha)^{1/2}.$$

Hence one of the two arcs in $\mathbb{T}$ between z_1 and z_3, or between z_2 and z_4 is mapped onto an arc of $\mathbb{T}$ of length less than $C_3\alpha^{1/2}$. Adding the arc between z_1 and z_4 we get, in all four cases, an arc $I \subset \mathbb{T}$ such that for the length $l(I)$ of I, we have $l(I) \geq 10\pi/9$ and

$$l(h(I)) < \alpha + C_3\alpha^{1/2} < C_4\alpha^{1/2}.$$

Since $|h(z_4)| = 1$,

$$\left|\int_{\mathbb{T}} h\,|\mathrm{d}z|\right| = \left|\int_I h\,|\mathrm{d}z| + \int_{\mathbb{T}\setminus I} h\,|\mathrm{d}z|\right| \geq \frac{10\pi}{9}(1 - C_4\alpha^{1/2}) - \frac{8\pi}{9}.$$

The first integral is 0 by the definition of $\widetilde{\varphi}$ and since $h(0) = 0$. Therefore (4.7) holds if $\alpha < C_5^{-1}$ where C_5 does not depend on z_0. Using Lemma 4.4.2 and (4.9) then shows that there exists a constant K such that

$$\left|\frac{\widetilde{\varphi}_{\overline{z}}(z_0)}{\widetilde{\varphi}_z(z_0)}\right| \leq K < 1$$

for $z_0 \in \mathbb{D}$. Therefore $\widetilde{\varphi}$ is quasiconformal in $\mathbb{D}$. □

5

The Beltrami differential equation

The aim of this chapter is to solve the Beltrami differential equation

$$f_{\bar{z}} = \mu f_z, \tag{5.1}$$

where μ is a measurable function on any Riemann surface with

$$||\mu||_\infty \leq k < 1$$

almost everywhere. The solution of the Beltrami differential equation is a quasiconformal mapping, and all quasiconformal mappings arise in this way, giving a correspondence between quasiconformal mappings and Beltrami differentials. We first tackle the case in $\mathbb{C}$ before moving onto Riemann surfaces in full generality. Further, we will see how quasiconformal maps vary as their corresponding Beltrami differentials vary. The material in this chapter is standard and can be found in other sources, for example [3], [27], or [32].

5.1 Integral transforms

The Cauchy transform P acts on functions $h \in L^p(\mathbb{C})$ for $p > 2$, and is defined by

$$Ph(w) = -\frac{1}{\pi} \int_{\mathbb{C}} h(z) \left(\frac{1}{z - w} - \frac{1}{z} \right) \mathrm{d}x \, \mathrm{d}y. \tag{5.2}$$

Lemma 5.1.1. *Ph is Hölder continuous, with Hölder exponent* $1 - 2/p$.

Proof. First we show that the integral in (5.2) is convergent.

$$|Ph(w)| = \frac{1}{\pi}\left|\int_{\mathbb{C}} h(z)\left(\frac{1}{z-w}-\frac{1}{z}\right) \mathrm{d}x\,\mathrm{d}y\right|$$
$$\le \frac{|w|}{\pi}\left(\int_{\mathbb{C}} |h(z)|^p \,\mathrm{d}x\,\mathrm{d}y\right)^{1/p}\left(\int_{\mathbb{C}} \frac{1}{(|z-w||z|)^q}\mathrm{d}x\,\mathrm{d}y\right)^{1/q}$$

using the Hölder inequality, where q is the conjugate exponent to p. Since $p > 2$, then $1 < q < 2$. Now

$$\int_{\mathbb{C}} \frac{1}{(|z(z-w)|)^q}\,\mathrm{d}x\,\mathrm{d}y = |w|^{2-2q}\int_{\mathbb{C}} \frac{1}{|z(z-1)|^q}\,\mathrm{d}x\,\mathrm{d}y$$

where the last integral is some constant K_p depending only on p. Hence

$$|Ph(w)| \le |w|^{1-2/p}K_p||h||_p. \tag{5.3}$$

Now if $\widetilde{h}(z) = h(z+w_1)$, then

$$P\widetilde{h}(w_2-w_1) = -\frac{1}{\pi}\int_{\mathbb{C}} h(z+w_1)\left(\frac{1}{z+w_1-w_2}-\frac{1}{z}\right)\,\mathrm{d}x\,\mathrm{d}y$$
$$= -\frac{1}{\pi}\int_{\mathbb{C}} h(z)\left(\frac{1}{z-w_2}-\frac{1}{z-w_1}\right)\,\mathrm{d}x\,\mathrm{d}y = Ph(w_2)-Ph(w_1).$$

Then by applying (5.3) to $\widetilde{h}$, and noting that $||\widetilde{h}||_p = ||h||_p$, we have that

$$|Ph(w_1)-Ph(w_2)| \le K_p||h||_p|w_1-w_2|^{1-2/p}.$$

□

The second integral operator is defined on functions in C_0^2, the space of twice continuously differentiable functions with compact support. The Hilbert transform T of $f \in C_0^2$ is the Cauchy principal value of

$$Th(w) = \lim_{\epsilon\to 0}\left(-\frac{1}{\pi}\int_{|z-w|>\epsilon}\frac{h(z)}{(z-w)^2}\,\mathrm{d}x\,\mathrm{d}y\right). \tag{5.4}$$

Lemma 5.1.2. *For $h \in C_0^2$, Th is well defined and $Th \in C^1$. Moreover,*

$$(Ph)_{\overline{z}} = h, \tag{5.5}$$
$$(Ph)_z = Th, \tag{5.6}$$
$$\int_{\mathbb{C}} |Th|^2\,\mathrm{d}x\,\mathrm{d}y = \int_{\mathbb{C}} |h|^2\,\mathrm{d}x\,\mathrm{d}y. \tag{5.7}$$

Proof. Initially, just assume that $h \in C_0^1$, which is enough for

$$(Ph)_{\overline{w}} = -\frac{1}{\pi}\int_{\mathbb{C}}\frac{h_{\overline{z}}}{z-w}\,\mathrm{d}x\,\mathrm{d}y,$$

and

$$(Ph)_w = -\frac{1}{\pi}\int_{\mathbb{C}} \frac{h_z}{z-w}\,\mathrm{d}x\,\mathrm{d}y.$$

Now let γ_ϵ be a circle with centre w and radius ϵ, and Ω_ϵ be $\mathbb{C}\backslash\{z : |z-w| < \epsilon\}$. By Stokes theorem, (3.1),

$$\begin{aligned}
-\frac{1}{\pi}\int_{\mathbb{C}} \frac{h_{\overline{z}}}{z-w}\,\mathrm{d}x\,\mathrm{d}y &= \frac{1}{2\pi i}\int_{\mathbb{C}} \frac{h_{\overline{z}}}{z-w}\,\mathrm{d}z\,\mathrm{d}\overline{z}\\
&= -\frac{1}{2\pi i}\int_{\mathbb{C}} \frac{\mathrm{d}h\,\mathrm{d}z}{z-w} = \lim_{\epsilon\to 0}\frac{1}{2\pi i}\int_{\gamma_\epsilon} \frac{h\,\mathrm{d}z}{z-w} = h(w),
\end{aligned}$$

and

$$\begin{aligned}
-\frac{1}{\pi}\int_{\mathbb{C}} \frac{h_z}{z-w}\,\mathrm{d}x\,\mathrm{d}y &= \frac{1}{2\pi i}\int_{\mathbb{C}} \frac{\mathrm{d}h\,\mathrm{d}\overline{z}}{z-w}\\
&= \lim_{\epsilon\to 0}\left(-\frac{1}{2\pi i}\int_{\gamma_\epsilon} \frac{h\,\mathrm{d}\overline{z}}{z-w} + \frac{1}{2\pi i}\int_{\Omega_\epsilon} \frac{h\,\mathrm{d}z\,\mathrm{d}\overline{z}}{(z-w)^2}\right) = Th(w).
\end{aligned}$$

This proves (5.5) and (5.6). Now, under the assumption that $h \in C_0^2$, by applying (5.6) to h_z and using the fact that

$$P(h_z) = Th - Th(0),$$

we have

$$(Th)_{\overline{z}} = P(h_z)_{\overline{z}} = h_z,$$

and

$$(Th)_z = P(h_z)_z = T(h_z) = P(h_{zz}) + Th_z(0),$$

which shows that $Th \in C^1$ and $Ph \in C^2$. Since h has compact support, then as $z \to \infty$, $Ph = O(1)$ and $Th = O(|z|^{-2})$, which is enough to justify the following calculation to verify that T is an isometry.

$$\begin{aligned}
\int_{\mathbb{C}} |Th|^2\,\mathrm{d}x\,\mathrm{d}y &= -\frac{1}{2i}\int_{\mathbb{C}} (Ph)_z(\overline{Ph})_{\overline{z}}\,\mathrm{d}z\,\mathrm{d}\overline{z}\\
&= \frac{1}{2i}\int_{\mathbb{C}} Ph(\overline{Ph})_{\overline{z}z}\,\mathrm{d}z\,\mathrm{d}\overline{z} = \frac{1}{2i}\int_{\mathbb{C}} (Ph)\overline{h}_{\overline{z}}\,\mathrm{d}z\,\mathrm{d}\overline{z}\\
&= -\frac{1}{2i}\int_{\mathbb{C}} \overline{h}(Ph)_{\overline{z}}\,\mathrm{d}z\,\mathrm{d}\overline{z} = \int_{\mathbb{C}} |h|^2\,\mathrm{d}x\,\mathrm{d}y.
\end{aligned}$$

□

Since C_0^2 is dense in L^2, then T can be extended to an isometry on the whole of L^2 by continuity. However, P cannot be extended in the same way, since the integral defining P is meaningless for $h \in L^2$. The solution to this problem is the Calderon–Zygmund inequality, which in effect says that the isometry relation

(5.7) can be replaced by

$$||Th||_p \leq C_p||h||_p, \tag{5.8}$$

for any $p > 1$. This allows us to extend T to L^p, and for $p > 2$ the integral for P is well defined. The proof of the Calderon–Zygmund inequality is omitted, but can be found in [3], for example.

Lemma 5.1.3. *For $h \in L^p$ with $p > 2$, the equations*

$$(Ph)_{\bar{z}} = h$$
$$(Ph)_z = Th$$

hold in the sense of distributions.

Proof. If the two equations are to hold in the sense of distributions, then for all test functions $\varphi \in C_0^1$,

$$\int_{\mathbb{C}} (Ph)\varphi_{\bar{z}} = -\int_{\mathbb{C}} \varphi h$$

and

$$\int_{\mathbb{C}} (Ph)\varphi_z = -\int_{\mathbb{C}} \varphi Th.$$

The equations hold when $h \in C_0^2$, and we can approximate h in the L^p-norm by $h_n \in C_0^2$. Since

$$||Th - Th_n||_p \leq C_p||h - h_n||_p$$

and from Lemma 5.1.1

$$|P(h - h_n)| \leq K_p||h - h_n||_p|z|^{1-2/p}$$

then the result holds since φ has compact support. □

5.2 Solution of the Beltrami equation

Initially we will deal with the case where μ in (5.1) has compact support, and $||\mu||_\infty \leq k < 1$. This means that the solution f will be analytic at ∞. Now fix $p > 2$ so that $kC_p < 1$, where C_p is the constant from the Calderon–Zygmund inequality (5.8).

Theorem 5.2.1. *Under the above assumptions, there is a unique solution of the Beltrami equation $f_{\bar{z}} = \mu f_z$ such that $f(0) = 0$ and $f_z - 1 \in L^p$.*

Proof. First we prove the uniqueness assertion. Assume f is a solution, then $f_{\bar{z}} = \mu f_z \in L^p$ and $P(f_{\bar{z}})$ exists. Then the function

$$F = f - P(f_{\bar{z}})$$

satisfies $F_{\overline{z}} = 0$ in the sense of distributions. By Weyl's lemma, F is analytic. Since $f_z - 1 \in L^p$, then $F' - 1 \in L^p$, which can only be satisfied if $F' = 1$, and the normalization $f(0) = 0$ implies that $F(z) = z$. Thus $f = P(f_{\overline{z}}) + z$, and $f_z = T(\mu f_z) + 1$.

If g is another solution, then

$$f_z - g_z = T(\mu(f_z - g_z)),$$

and by the Calderon–Zygmund inequality,

$$||f_z - g_z||_p \leq kC_p||f_z - g_z||_p.$$

Since we are assuming that $kC_p < 1$, then $f_z = g_z$ almost everywhere, and, via the Beltrami equation, $f_{\overline{z}} = g_{\overline{z}}$ almost everywhere. Therefore, $f - g$ and $\overline{f} - \overline{g}$ are both analytic, and the difference must be a constant. The normalization now shows that $f = g$.

Moving now onto the existence of a solution, consider the equation

$$h = T(\mu h) + T\mu.$$

The linear operator $h \to T(\mu h)$ on L^p has norm $\leq kC_p < 1$, which means that the series

$$h = T\mu + T\mu T\mu + T\mu T\mu T\mu + \cdots \tag{5.9}$$

converges in L^p, and solves $h = T(\mu h) + T\mu$. If h is given by (5.9), then

$$f = P(\mu(h+1)) + z \tag{5.10}$$

is the solution of the Beltrami equation. Since μ has compact support, then $\mu(h+1) \in L^p$ and so $P(\mu(h+1))$ is well defined and continuous. Also,

$$f_{\overline{z}} = \mu(h+1)$$
$$f_z = T(\mu(h+1)) + 1 = h + 1$$

and $f_z - 1 = h \in L^p$. □

The function given by f in (5.10) is called the normal solution of the Beltrami equation.

Lemma 5.2.1. *Let μ_n, μ satisfy the hypothesis of Theorem 5.2.1 with uniformly bounded supports, and f_n, f be the respective normal solutions. Suppose $\mu_n \to \mu$ almost everywhere, then $||(f_n)_z - f_z||_p \to 0$ and $f_n \to f$ uniformly on compact sets.*

Proof. From the definition of h in Theorem 5.2.1,

$$||h||_p \leq kC_p||h||_p + C_p||\mu||_p$$

and so

$$||h||_p \leq \frac{C_p}{1 - kC_p}||\mu||_p.$$

Thus

$$||f_{\bar{z}}||_p \leq \frac{1}{1-kC_p}||\mu||_p.$$

Also, the Hölder condition on P gives

$$|f(w_1) - f(w_2)| \leq \frac{K_p}{1-kC_p}||\mu||_p|w_1 - w_2|^{1-2/p} + |w_1 - w_2|. \tag{5.11}$$

So, we have

$$\begin{aligned} ||(f_n)_z - f_z||_p &\leq ||T[\mu_n((f_n)_z - f_z)]||_p + ||T[(\mu_n - \mu)f_z]||_p \\ &\leq kC_p||(f_n)_z - f_z||_p + C_p||(\mu_n - \mu)f_z||_p, \end{aligned}$$

which proves the claim. □

To show that the normal solution f has derivatives if μ does, we require the following lemma, which we quote without proof.

Lemma 5.2.2. *If p and q are continuous and have locally integrable derivatives in the sense of distributions that satisfy $p_{\bar{z}} = q_z$, then there exists a function $f \in C^1$ such that $f_z = p$ and $f_{\bar{z}} = q$.*

Lemma 5.2.3. *If μ has a derivative in the sense of distributions and $\mu_z \in L^p$ for $p > 2$, then $f \in C^1$.*

Proof. The aim is to solve the system $f_z = \lambda$, $f_{\bar{z}} = \mu\lambda$ for λ. Lemma 5.2.2 says that this will be possible if

$$\lambda_{\bar{z}} = (\mu\lambda)_z = \lambda_z\mu + \lambda\mu_z$$

or

$$(\log\lambda)_{\bar{z}} = \mu(\log\lambda)_z + \mu_z. \tag{5.12}$$

Now the equation $q = T(\mu q) + T\mu_z$ can be solved for $q \in L^p$, and set

$$\sigma = P(\mu q + \mu_z) + C$$

where C is a constant such that $\sigma \to 0$ as $z \to \infty$. Then σ is continuous and

$$\begin{aligned} \sigma_{\bar{z}} &= \mu q + \mu_z, \\ \sigma_z &= T(\mu q + \mu_z) = q. \end{aligned}$$

Therefore $\lambda = e^\sigma$ solves (5.12), and so the system we required to be solved can indeed be solved for $f \in C^1$. Normalizing by $f(0) = 0$ gives us the normal solution, since as $z \to \infty$, $\sigma \to 0$, $\lambda \to 1$, and $f_z \to 1$.

The Jacobian

$$|f_z|^2 - |f_{\bar{z}}|^2 = (1 - |\mu|^2)e^{2\sigma}$$

is strictly positive, and hence the mapping f is locally one-to-one, and since $f(z) \to \infty$ as $z \to \infty$, f is a global homeomorphism. □

The inverse function f^{-1} is also K-quasiconformal with complex dilatation $\widetilde{\mu} = \mu_{f^{-1}}$ which satisfies $|\widetilde{\mu} \circ f| = |\mu|$. Now,

$$\begin{aligned}\int_{\mathbb{C}} |\widetilde{\mu}|^p \, \mathrm{d}\zeta \, \mathrm{d}\eta &= \int_{\mathbb{C}} |\mu|^p (|f_z|^2 - |f_{\overline{z}}|^2) \, \mathrm{d}x \, \mathrm{d}y \\ &\le \int_{\mathbb{C}} |\mu|^p |f_z|^2 \, \mathrm{d}x \, \mathrm{d}y = \int_{\mathbb{C}} |\mu|^{p-2} |f_{\overline{z}}|^2 \, \mathrm{d}x \, \mathrm{d}y \\ &\le \left(\int_{\mathbb{C}} |\mu|^p \, \mathrm{d}x \, \mathrm{d}y \right)^{(p-2)/p} \left(\int_{\mathbb{C}} |f_{\overline{z}}|^p \, \mathrm{d}x \, \mathrm{d}y \right)^{2/p}\end{aligned}$$

and so

$$||\widetilde{\mu}||_p \le (1 - kC_p)^{-2/p} ||\mu||_p.$$

Applying the inequality (5.11) to the inverse function gives

$$|z_1 - z_2| \le K_p (1 - kC_p)^{-1-2/p} ||\mu||_p |f(z_1) - f(z_2)|^{1-2/p} + |f(z_1) - f(z_2)|. \tag{5.13}$$

Theorem 5.2.2. *For any μ with compact support and $||\mu||_\infty \le k < 1$, the normal solution to the Beltrami equation is a quasiconformal homeomorphism which satisfies $\mu_f = \mu$.*

Proof. We can find a sequence of functions $\mu \in C^1$ which satisfy $\mu_n \to \mu$ almost everywhere, $|\mu_n| \le k$, and $\mu_n = 0$ outside some fixed circle. The corresponding normal solutions f_n satisfy inequality (5.13), and because we have $f_n \to f$ and $||\mu_n||_p \to ||\mu||_p$, then f also satisfies (5.13) and so f is injective. Since f is a uniform limit of K-quasiconformal mappings, it itself is a K-quasiconformal mapping, and so has locally integrable partial derivatives in the sense of distributions. Also $f_z \ne 0$ almost everywhere, so μ_f is defined almost everywhere, and is equal to μ. □

Exercise 5.2.1. Using the Hölder inequality, find another proof of Theorem 3.1.6, that is, that f maps sets of measure zero onto sets of measure zero.

We shall now drop the assumption that μ has compact support.

Theorem 5.2.3. *For any $\mu \in L^\infty(\mathbb{C})$ with $||\mu||_\infty < 1$, there exists a unique quasiconformal mapping f^μ with complex dilatation μ, which is normalized so that f^μ fixes $0, 1$ and ∞.*

Proof. If μ has compact support, then by Theorem 5.2.2, f only needs to be normalized. Now assume $\mu = 0$ in a neighbourhood of 0, and set

$$\widetilde{\mu}(z) = \mu\left(\frac{1}{z}\right) \frac{z^2}{\overline{z}^2}.$$

Then $\widetilde{\mu}$ has compact support, and it can be verified that

$$f^\mu(z) = \frac{1}{f^{\widetilde{\mu}}(1/z)}.$$

For the general case, write $\mu = \mu_1 + \mu_2$ where $\mu_1 = 0$ in a neighbourhood of ∞, and $\mu_2 = 0$ in a neighbourhood of 0. We want to find λ so that $f^\lambda \circ f^{\mu_2} = f^\mu$ and $f^\lambda = f^\mu \circ (f^{\mu_2})^{-1}$. This is achieved for

$$\lambda = \left(\left(\frac{\mu - \mu_2}{1 - \mu\overline{\mu_2}} \right) \left(\frac{f_z^{\mu_2}}{\overline{f_z^{\mu_2}}} \right) \right) \circ (f^{\mu_2})^{-1},$$

and λ has compact support. □

Theorem 5.2.4. *Let M be a Riemann surface, and μ a $(-1, 1)$ differential form such that $||\mu||_\infty \leq k < 1$. Then there exists a Riemann surface N, and a $(1 + k)/1 - k$-quasiconformal homeomorphism $f^\mu : M \to N$ such that $\mu_f = \mu$.*

Proof. If g is a map between charts where μ_1 and μ_2 are the representations of the Beltrami differential μ, then

$$(\mu_2 \circ g)\frac{\overline{g'}}{g'} = \mu_1.$$

Thus $|\mu| : M \to \mathbb{R}$ is a well-defined function on M, and $||\mu||_\infty$ makes sense.

We consider the different cases of Riemann surfaces as classified by the Riemann mapping theorem. The cases where M is S^2, $\mathbb{C}$, or $\mathbb{C}$ with one or two points removed, are covered by Theorem 5.2.3. If M is a torus, there is a projection $\pi : \mathbb{C} \to M$ and a group $\Gamma \simeq \mathbb{Z} \oplus \mathbb{Z}$ that acts on $\mathbb{C}$ such that $M \simeq \mathbb{C}/\Gamma$. Let $\Gamma =< T_1, T_2 >$, where T_1, T_2 are translations of the plane and are linearly independent over $\mathbb{R}$. Then lifting μ to $\mathbb{C}$ via

$$\mu(w) = \widetilde{\mu}(\pi(z))\frac{\overline{\pi'(z)}}{\pi'(z)}$$

gives an L^∞ function $\widetilde{\mu}$, and then solve for $f^{\widetilde{\mu}} : \mathbb{C} \to \mathbb{C}$. Now

$$\widetilde{\mu} \circ T_1 = \widetilde{\mu} = \widetilde{\mu} \circ T_2,$$

and if $T \in \Gamma$, then $(f^{\widetilde{\mu}})^{-1} \circ T \circ f^{\widetilde{\mu}}$ is a conformal map of $\mathbb{C}$ with no fixed points, so must be a translation, that is, it is in $\Gamma' =< T_1', T_2' >$, where

$$T_i' = (f^{\widetilde{\mu}})^{-1} \circ T_i \circ f^{\widetilde{\mu}}$$

for $i = 1, 2$. Then $\mathbb{C}/\Gamma' \simeq N$, which is another torus, and so f^μ, which lifts to $\widetilde{f^\mu}$, is a quasiconformal homeomorphism between tori.

Finally, if M is any other Riemann surface, then by the Riemann mapping theorem, it has the unit disk $\mathbb{D}$ as its universal cover. We have $M \simeq \mathbb{D}/\Gamma$, where Γ is a group of Möbius transformations of $\mathbb{D}$. Now μ is a Beltrami differential on M and, as in the case of the torus, can be lifted to an L^∞ function $\widetilde{\mu}$ on $\mathbb{D}$

which satisfies

$$(\widetilde{\mu} \circ A)\frac{\overline{A'}}{A'} = \mu$$

for any $A \in \Gamma$. Extend $\widetilde{\mu}$ to $\mathbb{C}$ by $\overline{\widetilde{\mu}(z^*)} = \widetilde{\mu}(z)$, where z^* denotes reflection in $\partial\mathbb{T}$. Then by Theorem 5.2.3, $f^{\widetilde{\mu}} : \mathbb{C} \to \mathbb{C}$ and, by symmetry, $f^{\widetilde{\mu}} : \mathbb{D} \to \mathbb{D}$. If $A \in \Gamma$, $(f^{\widetilde{\mu}})^{-1} \circ A \circ f^{\widetilde{\mu}}$ is a conformal map of $\mathbb{D}$ to itself, and hence must be a Möbius transformation. Thus

$$(f^{\widetilde{\mu}})^{-1} \circ \Gamma \circ f^{\widetilde{\mu}} = \Gamma',$$

another group of Möbius transformations and define $N \simeq \mathbb{D}/\Gamma'$. We can conclude that $f^{\mu} : M \to N$, where f^{μ} lifts to $f^{\widetilde{\mu}}$. □

Let f be a quasiconformal mapping with maximal dilatation K, and assume that we can write $f = f_2 \circ f_1$, where f_1 and f_2 are K_1, K_2 quasiconformal respectively. Clearly $K \leq K_1 K_2$, but using Theorem 5.2.3 we can find f_1 and f_2 so that $K = K_1 K_2$.

Theorem 5.2.5. *Let f be a quasiconformal mapping with maximal dilatation K, and $0 < t < 1$. Then $f = f_2 \circ f_1$, where f_1 is K^t-quasiconformal, and f_2 is K^{1-t}-quasiconformal.*

Proof. Let μ be the complex dilatation of f, and choose the complex dilatation μ_1 of f_1 in the following way. Let $\mu_1(z)$ be the point on the line segment from 0 to $\mu(z)$ such that $d_h(0, \mu_1(z)) = t d_h(0, \mu(z))$, where d_h is the hyperbolic distance on $\mathbb{D}$. Then

$$\frac{1 + |\mu_1(z)|}{1 - |\mu_1(z)|} = \left(\frac{1 + |\mu(z)|}{1 - |\mu(z)|}\right)^t$$

which shows that f_1 is K^t-quasiconformal. If μ_2 is the complex dilatation of $f_2 = f \circ f_1^{-1}$, and $w = f_1(z)$, then

$$|\mu_2(w)| = \left| \frac{\mu(z) - \mu_1(z)}{1 - \overline{\mu_1(z)}\mu(z)} \right|,$$

and so

$$\log\left(\frac{1 + |\mu_2(w)|}{1 - |\mu_2(w)|}\right) = 2d_h(\mu_1(z), \mu(z))$$

$$= 2(1-t)d_h(0, \mu(z)) = (1-t)\log\left(\frac{1 + |\mu(z)|}{1 - |\mu(z)|}\right),$$

and so f_2 is K^{1-t}-quasiconformal. □

Remark. From Theorem 5.2.5, we can conclude that if f is K-quasiconformal and $\epsilon > 0$, then we can decompose f as

$$f = f_n \circ \cdots \circ f_1$$

where each map f_i for $i = 1, \ldots, n$ is $(1 + \epsilon)$-quasiconformal.

As a further consequence of Theorem 5.2.3, we can show that every quasiconformal map from $\mathbb{D}$ to a Jordan domain extends to a homeomorphism of the closures.

Proposition 5.2.1. *Let Ω be a Jordan domain, and $f : \mathbb{D} \to \Omega$ quasiconformal. Then f extends to a homeomorphism of $\overline{\mathbb{D}}$ onto $\overline{\Omega}$.*

Proof. For a given quasiconformal map $f : \mathbb{D} \to \Omega$, let $\mu = \mu_f$ be its Beltrami coefficient. Extend μ to an element of $L^\infty(\mathbb{C})$ by setting $\mu = 0$ on $\mathbb{C} \setminus \mathbb{D}$. By Theorem 5.2.3 there is a unique normalized quasiconformal map f^μ of $\widehat{\mathbb{C}}$ with Beltrami coefficient μ.

Now let $g = f^\mu \circ f^{-1}$. Then g is a quasiconformal mapping of $\mathbb{D}$ and by formula (1.6), $\mu_g = 0$ almost everywhere on $\mathbb{D}$. Therefore g is a conformal mapping of $\mathbb{D}$. Since $f^\mu(\mathbb{D})$ is a Jordan domain, then Caratheodory's theorem from complex analysis says that g extends to a homeomorphism of $\overline{\mathbb{D}}$ onto $\overline{f^\mu(\mathbb{D})}$. Since we have $f = g^{-1} \circ f^\mu$, this completes the proof. □

Exercise 5.2.2. Using Theorem 5.2.3, prove that there is no quasiconformal mapping of $\mathbb{D}$ onto $\mathbb{C}$. (Hint: use Liouville's theorem, the corresponding result for conformal maps.)

5.3 Dependence on Beltrami coefficients

In this section, we explore how the normalized quasiconformal map f^μ of Theorem 5.2.3 depends on its Beltrami coefficient μ. Lemma 5.2.1 shows that the normal solution, which will be denoted in this section by F^μ, depends continuously on μ. Let

$$||f||_{p,R} = \left(\int_{\{|z| \le R\}} |f|^p \, dx \, dy \right)^{1/p} .$$

Lemma 5.3.1. *For $p > 2$ and $\mu_n \in L^\infty(\mathbb{C})$ with $||\mu_n||_\infty < 1$ for all n, and if $||\mu_n||_\infty \to 0$ as $n \to \infty$, then*

$$||(f^\mu)_z - 1||_{p,1} \to 0.$$

Proof. Assume that the supports of the μ_n are uniformly bounded and fix $p > 2$. Since $||\mu_n||_\infty C_p < 1$ for sufficiently small $||\mu_n||_\infty$, then for these μ_n, there exists a normal solution F^{μ_n}. The normal solution F^μ and the normalized solution f^μ are related by

$$f^\mu(z) = \frac{F^\mu(z)}{F^\mu(1)}.$$

Since $F^{\mu_n} \to 1$ as $||\mu_n||_\infty \to 0$, then Lemma 5.2.1 implies

$$||(f_n^\mu)_z - 1||_{p,1} \to 0$$

as $n \to \infty$. Now set

$$\widetilde{f^\mu}(z) = \frac{1}{f^\mu(1/z)}$$

and

$$\widetilde{\mu}(z) = \mu_{\widetilde{f^\mu}}(z) = \mu\left(\frac{1}{z}\right)\frac{z^2}{\overline{z}^2}.$$

Then $\widetilde{f^\mu}(z)$ is the normalized quasiconformal mapping $f^{\widetilde{\mu}}$ with Beltrami coefficient $\widetilde{\mu}$. Observe that the supports of the μ_n are bounded if and only if the $\widetilde{\mu_n}$ vanish on some fixed neighbourhood of 0.

Suppose that all the $\widetilde{\mu_n}$ vanish on a fixed neighbourhood of 0. Then it will be shown that

$$||(\widetilde{f^{\mu_n}})_z - 1||_{p,2} \to 0 \tag{5.14}$$

as $||\widetilde{\mu_n}||_\infty \to 0$. Since

$$\widetilde{f^{\mu_n}}(z) = \frac{F^{\mu_n}(1)}{F^{\mu_n}(\frac{1}{z})}$$

and $F^{\mu_n}(1) \to 1$, it is enough to estimate

$$I(\mu_n) := \int_{\{|z|<2\}} |(\widetilde{F}^{\mu_n})_z - 1|^p \, dx\, dy,$$

where

$$\widetilde{F}^{\mu_n}(z) = \left(F^{\mu_n}\left(\frac{1}{z}\right)\right)^{-1}.$$

Divide the domain of integration in the following way:

$$I(\mu_n) = \left(\int_{\{1/2<|z|<R\}} + \int_{\{|z|\geq R\}}\right) \left| \frac{z^2 (F^{\mu_n})_z(z)}{(F^{\mu_n}(z))^2} - 1\right|^p \frac{dx\, dy}{|z|^4} = I_1 + I_2.$$

For large enough R, every F^{μ_n} is holomorphic on $\{|z| > R\}$, and F^{μ_n} converges to $F^0(z) = z$ uniformly on $\{|z| > R\}$, and therefore $I_2 \to 0$ as $n \to \infty$. Writing I_1 as

$$I_1 = \int_{\{1/2<||z|<R\}} \left| \frac{z^2((F^{\mu_n})_z(z) - 1)}{(F^\mu(z))^2} + \frac{z^2}{(F^\mu(z))^2} - 1\right|^p \frac{dx\, dy}{|z|^4},$$

and using Lemma 5.2.1 shows that $I_1 \to 0$ as $n \to \infty$.

For a general μ, set $\nu(z) = \mu(z)$ on $\mathbb{D}$ and 0 elsewhere. Let f^ν be the normalized quasiconformal mapping with Beltrami coefficient ν and then set $g^\mu = f^\mu \circ (f^\nu)^{-1}$. Again by Lemma 5.2.1, f^{ν_n} converges to the identity uniformly on $\mathbb{C}$ as $n \to \infty$. Hence we can assume that for all n, $f^{\nu_n}(\mathbb{D})$ is contained in $\{|z| < 2\}$ and contains $\{|z| \leq 1/2\}$. Then every $\mu_{g^{\mu_n}}$ vanishes on some fixed neighbourhood of 0. Since

$$(f^{\mu_n})_z = ((g^{\mu_n})_z \circ f^{\nu_n})(f^{\nu_n})_z + ((g^{\mu_n})_{\overline{z}} \circ f^{\nu_n})\overline{(f^{\nu_n})_z}$$

and since either $(f^{\nu_n})_{\overline{z}} = 0$ or $(g^{\mu_n})_{\overline{z}} \circ f^{\nu_n} = 0$ almost everywhere on $\mathbb{C}$, then

$$||(f^{\mu_n})_z - 1||_{p,1} \leq ||(((g^{\mu_n})_z - 1) \circ f^{\nu_n})(f^{\nu_n})_z||_{p,1} + ||(f^{\nu_n})_z - 1||_{p,1}.$$

We have already shown that $||(f^{\nu_n})_z - 1||_{p,1} \to 0$ as $n \to \infty$, and

$$\begin{aligned} &||(((g^{\mu_n})_z - 1) \circ f^{\nu_n})(f^{\nu_n})_z||_{p,1}^p \\ &\leq \frac{1}{1-k^2} \int_{f^{\nu_n}(\mathbb{D})} |(g^{\mu_n})_z - 1|^p |(f^{\nu_n})_z \circ (f^{\nu_n})^{-1}|^{p-2} \, dx \, dy \\ &\leq \frac{1}{1-k^2} \left(\int_{\{|z|<2\}} |(g^{\mu_n})_z - 1|^{2p} \, dx \, dy \int_{\mathbb{D}} |(f^{\nu_n})_z|^{2p-2} \, dx \, dy \right)^{1/2} \end{aligned}$$

where $k = ||\mu_n||_\infty$. Using (5.14) by replacing $\widetilde{f}^{\mu_n}$ and p with g^{μ_n} and $2p$ respectively,

$$||(((g^{\mu_n})_z - 1) \circ f^{\nu_n})(f^{\nu_n})_z||_{p,1}^p \to 0$$

as $n \to \infty$, which proves the lemma. □

We quote without proof the following lemma which gives an integral representation for the normalized quasiconformal map f^μ. See [27] for a proof.

Lemma 5.3.2. *Fix $p > 8$ arbitrarily, and let $\mu \in L^\infty$ such that $||\mu||_\infty < 1$ and $||\mu||_\infty C_p < 1$. Then the normalized quasiconformal mapping f^μ satisfies*

$$\begin{aligned} f^\mu(w) = w &- \frac{1}{\pi} \int_{\mathbb{D}} (f^\mu)_z(z) \left(\frac{1}{z-w} - \frac{w}{z-1} + \frac{w-1}{z} \right) dx \, dy \qquad (5.15) \\ &- \frac{1}{\pi} \int_{\mathbb{D}} \frac{(\widetilde{f}^\mu)_z(z)}{(\widetilde{f}^\mu(z))^2} \left(\frac{w^2 z}{1-zw} - \frac{wz}{1-z} \right) dx \, dy \end{aligned}$$

for all $w \in \mathbb{D}$, and where $\widetilde{f}^\mu(z) = 1/f^\mu(1/z)$.

Proposition 5.3.1. *If $\mu_n \to 0$ in the unit ball of L^∞, then the normalized quasiconformal mappings f^{μ_n} converge to the identity mapping uniformly on compact subsets of $\mathbb{C}$.*

Proof. First fix $p > 8$, and since $||\mu_n||_\infty \to 0$, we can assume that $||\mu_n||_\infty C_p < 1$ for all n. This means that the formula (5.15) is valid. Writing the right-hand side of (5.15) as $w + I(w) + \widetilde{I}(w)$, we will show that $I(w)$ and $\widetilde{I}(w)$ converge to 0.

Since $(f^{\mu_n})_{\overline{z}} = \mu_n((f^{\mu_n})_z - 1) + \mu_n$, using Hölder's inequality gives a constant M such that for $w \in \mathbb{D}$ and for all n,

$$|I(w)| \leq M(||(f^{\mu_n})_z - 1||_{p,1} + 1)||\mu_n||_\infty.$$

By Lemma 5.3.1, $I(w) \to 0$ uniformly on $\mathbb{D}$ as $||\mu_n||_\infty \to 0$.

It can be shown that there exists a neighbourhood U of the origin and $m > 0$ such that for $z \in U$,

$$|\widetilde{f}^\mu(z)| \geq m|z|^{p/(p-2)} \qquad (5.16)$$

and U and m can be chosen uniformly with respect to u_n. Therefore there is a constant M such that

$$|\widetilde{I}(w)| \leq M(||(\widetilde{f}^{\mu_n})_z - 1||_{p,1} + 1)||\widetilde{\mu_n}||_\infty,$$

where $\widetilde{\mu}(z)$ is defined as in the proof of Lemma 5.3.1. Since $\widetilde{f}^{\mu_n}$ is the normalized quasiconformal mapping for the Beltrami coefficient $\widetilde{\mu_n}$, and also since we have $||\mu_n||_\infty = ||\widetilde{\mu_n}||_\infty$, then Lemma 5.3.1 shows that $\widetilde{I}(w) \to 0$ uniformly on $\mathbb{D}$ as $||\mu_n||_\infty \to 0$. This shows that f^{μ_n} converges to the identity uniformly on $\mathbb{D}$ as $||\mu_n||_\infty \to 0$.

To finish off the proof, for every $0 < r < 1$, set $\mu_{n,r}(z) = \mu_n(z/r)$ and consider the normalized quasiconformal mapping $f^{\mu_{n,r}}$. Let

$$f_{n,r}(z) = \frac{f^{\mu_{n,r}}(rz)}{f^{\mu_{n,r}}(r)}.$$

Then since $\mu_{f_{n,r}}(z) = \mu_{n,r}(rz) = \mu_n(z)$ almost everywhere, and since $f_{n,r}$ satisfies the normalization conditions, then the uniqueness of solutions of the Beltrami equation shows that $f_{n,r} = f^{\mu_n}$. However, since $f^{\mu_{n,r}}$ converges to the identity uniformly on $\mathbb{D}$ as shown above, then so does $f_{n,r}$ on $\{|z| < 1/r\}$. Since r is arbitrary, f^{μ_n} converges to the identity locally uniformly on $\mathbb{C}$. □

Theorem 5.3.1. *Let $\{\mu(t)\}$ be a family of Beltrami coefficients depending on a real or complex parameter t. Suppose that $||\mu(t)||_\infty \to 0$ as $t \to 0$, and that $\mu(t)$ is differentiable at $t = 0$, that is, $\mu(t)$ can be written in the form*

$$\mu(t)(z) = t\nu(z) + t\epsilon(t)(z),$$

for $z \in \mathbb{C}$ and where $\nu, \epsilon(t) \in L^\infty$ such that $||\epsilon(t)||_\infty \to 0$ as $t \to 0$. Then

$$\dot{f}[\nu](w) = \lim_{t\to 0} \frac{f^{\mu(t)}(w) - w}{t}$$

exists for every $w \in \mathbb{C}$, and the convergence is uniform on compact subsets of $\mathbb{C}$. Moreover, $\dot{f}[\nu]$ has the integral representation

$$\dot{f}[\nu](w) = -\frac{1}{\pi}\int_{\mathbb{C}} \nu(z)\frac{w(w-1)}{z(z-1)(z-w)}\, dx\, dy \tag{5.17}$$

for $w \in \mathbb{C}$.

Proof. As for the proof of Proposition 5.3.1, formula (5.15) holds with $\mu = \mu(t)$ for every t. Express the right-hand side of (5.15) as $w + I_t(w) + \widetilde{I}_t(w)$. Since

$$(f^{\mu(t)})_{\overline{z}} = \mu(t)((f^{\mu(t)})_z - 1) + \mu(t)$$

and $||\mu(t)/t - \nu||_\infty \to 0$ as $t \to 0$, then by a similar argument to that in Proposition 5.3.1,

$$\lim_{t\to 0}\frac{I_t(w)}{t} = -\frac{1}{\pi}\int_{\mathbb{D}} \nu(z)\left(\frac{1}{z-w} - \frac{w}{z-1} + \frac{w-1}{z}\right) dx\, dy \tag{5.18}$$

uniformly on $\mathbb{D}$. Now set

$$\widetilde{\mu}(t)(z) = \mu_{\widetilde{f^\mu}}(z) = \mu(t)\left(\frac{1}{z}\right)\frac{z^2}{\overline{z}^2}$$

and

$$\widetilde{\nu}(z) = \nu\left(\frac{1}{z}\right)\frac{z^2}{\overline{z}^2}.$$

Since $||\widetilde{\mu}(t)/t - \widetilde{\nu}||_\infty \to 0$ as $t \to 0$, then again by a similar argument to that of Proposition 5.3.1,

$$\lim_{t\to 0}\frac{\widetilde{I}_t(w)}{t} = -\frac{1}{\pi}\int_{\mathbb{D}}\frac{\widetilde{\nu}(z)}{(\widetilde{f}^{\mu(t)}(z))^2}\left(\frac{w^2 z}{1-zw} - \frac{wz}{1-z}\right)\,dx\,dy$$

uniformly on $\mathbb{D}$. The estimate in (5.16) implies that the integrand of this integral is bounded by

$$M|z|^{-2p/(p-2)}\frac{|z|}{(1-|w|)|1-z|}$$

for $w \in \mathbb{D}$, for every t with a suitable large enough constant M. This bound is absolutely integrable on $\mathbb{D}$ for every $w \in \mathbb{D}$, and so $\widetilde{I}_t(w)/t$ converges to

$$-\frac{1}{\pi}\int_{\mathbb{D}}\nu\left(\frac{1}{z}\right)\frac{1}{\overline{z}^2}\left(\frac{w^2 z}{1-zw} - \frac{wz}{1-z}\right)\,dx\,dy$$

uniformly on compact subsets as $t \to 0$. Changing the variable z to $1/z$ in this integral and adding this to (5.18) gives (5.17) for every $w \in \mathbb{D}$. Finally, the same argument as the last part of Proposition 5.3.1 shows that $(f^{\mu(t)}(w) - w)/t$ converges to the right-hand side of (5.17) uniformly on compact subsets of $\mathbb{C}$ as $t \to 0$. □

Corollary 5.3.1. *Let $\{\mu(t)\}$ be a family of Beltrami coefficients depending on a real or complex parameter t. Suppose also that $\mu(t)$ is differentiable at $t = 0$, that is, $\mu(t)$ can be written in the form*

$$\mu(t)(z) = \mu(z) + t\nu(z) + t\epsilon(t)(z)$$

for $z \in \mathbb{C}$, with suitable μ in the unit ball of $L^\infty(\mathbb{C})$, $\nu, \epsilon(t) \in L^\infty(\mathbb{C})$ such that $||\epsilon(t)||_\infty \to 0$ as $t \to 0$. Then

$$f^{\mu(t)}(w) = f^\mu(w) + t\dot{f}^\mu[\nu](w) + o(|t|)$$

locally uniformly on $\mathbb{C}$ as $t \to 0$, for $w \in \mathbb{C}$, and where

$$\dot{f}^\mu[\nu](w) = -\frac{1}{\pi}\int_{\mathbb{C}}\nu(z)\frac{f^\mu(w)(f^\mu(w)-1)((f^\mu)_z(z))^2}{f^\mu(z)(f^\mu(z)-1)(f^\mu(z)-f^\mu(w))}\,dx\,dy.$$

Proof. Set $f_t = f^{\mu(t)} \circ (f^\mu)^{-1}$. Then the complex dilatation $\lambda(t)$ of f_t is given by

$$\lambda(t) = \mu_{ft} = \left(\frac{\mu(t) - \mu}{1 - \overline{\mu}\mu} \frac{(f^\mu)_z}{\overline{(f^\mu)_z}} \right) \circ (f^\mu)^{-1}.$$

Therefore $\lambda(t)$ is written as

$$\lambda(t) = t\dot{\lambda} + o(|t|)$$

in $L^\infty(\mathbb{C})$ as $t \to 0$, where

$$\dot{\lambda} = \left(\frac{\nu}{1 - |\mu|^2} \frac{(f^\mu)_z}{\overline{(f^\mu)_z}} \right) \circ (f^\mu)^{-1}.$$

Apply Theorem 5.3.1 to this family $\{f_t\}$, and conclude that $(f_t(w) - w)/t$ converges to

$$\dot{f}[\dot{\lambda}](w) = -\frac{1}{\pi} \int_{\mathbb{C}} \dot{\lambda}(z) \frac{w(w-1)}{z(z-1)(z-w)} \, dx \, dy$$

uniformly on compact subsets of $\mathbb{C}$ as $t \to 0$. Changing the variable in this integral from z to $(f^\mu)^{-1}(z)$ and noting that

$$\frac{f^{\mu(t)} - f^\mu}{t} = \left(\frac{f_t - f_0}{t} \right) \circ f^\mu$$

proves the corollary. □

Remark. If μ_t depends holomorphically on the complex parameter t, then $f^{\mu(t)}(w)$ is holomorphic with respect to t for every fixed $w \in \mathbb{C}$. Now fix $R > 0$, and set

$$\begin{aligned} ||f||_{B_{R,p}} = & \sup_{|z_1|,|z_2| \le R} \frac{|f(z_1) - f(z_2)|}{|z_1 - z_2|^{1-2/p}} \\ & + \left(\int_{|z| \le R} |f_z|^p \, dx \, dy \right)^{1/p} + \left(\int_{|z| \le R} |f_{\overline{z}}|^p \, dx \, dy \right)^{1/p} \end{aligned}$$

for every normalized quasiconformal mapping f of $\mathbb{C}$ and $p > 2$. Fix $0 \le k < 1$ such that $kC_p < 1$, then

$$\{||f^\mu||_{B_{R,p}} : ||\mu||_\infty \le k\}$$

is a bounded set. Now let $\{\mu_n\}_{n=1}^\infty$ be a sequence of Beltrami coefficients such that $||\mu_n||_\infty \le k$ for every n and $\mu_n \to \mu$ almost everywhere on $\mathbb{C}$ as $n \to \infty$. Then

$$||f^{\mu_n} - f^\mu||_{B_{R,p}} \to 0$$

as $n \to \infty$. In particular, f^{μ_n} converges to f^μ uniformly on compact subsets of $\mathbb{C}$ as $n \to \infty$.

If $\mu(t)$ is differentiable at $t = 0$ with respect to a real or complex parameter t, then

$$\lim_{t \to 0} \left|\left| \frac{f^{\mu(t)} - f^{\mu(0)}}{t} - (\dot{f}^{\mu(0)}) \left\{ \frac{\mathrm{d}\mu}{\mathrm{d}t}(0) \right\} \right|\right|_{B_{R,p}} = 0.$$

If $\mu(t)$ depends on t holomorphically, then $f^{\mu(t)}$ depends on t holomorphically with respect to the norm $||.||_{B_{R,p}}$.

6

Holomorphic motions and applications

In this chapter, we investigate holomorphic motions and their relationship to quasiconformal mappings. The main problem in the theory of holomorphic motions is that of extending them, and we show how this can be achieved in an equivariant way. We also show how holomorphic motions can be applied to Astala's theorem on area distortion in the theory of quasiconformal mappings.

6.1 Holomorphic motions

We saw at the end of the previous chapter that if $\mu(t)$ depends holomorphically on t, then $f^{\mu(t)}$ also depends holomorphically on t. The following definition was introduced by Mane, Sullivan, and Sad in [34].

Definition 6.1.1. *Let E be a subset of $\widehat{\mathbb{C}} = \mathbb{C} \cup \{\infty\}$. A holomorphic motion of E over $\mathbb{D}$ is a function $f : \mathbb{D} \times E \to \widehat{\mathbb{C}}$ satisfying*

(1) *$f(0, z) = z$ for all $z \in E$,*

(2) *$f(., z)$ is a holomorphic $\widehat{\mathbb{C}}$-valued function on $\mathbb{D}$ for each $z \in E$,*

(3) *$f(t, .)$ is injective on E for each $t \in \mathbb{D}$.*

Example 6.1.1. If $\mu \in L^\infty(\widehat{\mathbb{C}})$, $||\mu||_\infty \leq k < 1$, $t \in \mathbb{D}$, then $f(t, z) = f^{t\mu}(z)$ is a holomorphic motion of $\widehat{\mathbb{C}}$.

Theorem 6.1.1. *A holomorphic motion $f(t, z)$ of $E = \mathbb{D}$ can be extended to a homeomorphism, in the z variable, of $\overline{\mathbb{D}}$ onto a closed quasidisk in $\mathbb{C}$, for each $t \in \mathbb{D}$.*

Recall from the comments following Definition 2.3.1 that a quasidisk is the image of a disk under a quasiconformal mapping. The proof of Theorem 6.1.1

relies on the following Schottky theorem, which is proved in [42] and is omitted here.

Theorem 6.1.2. *If h is analytic on $\mathbb{D}$ and not equal to 0 or 1, then*

$$|h(z)| \leq \Phi\left(|h(0)|, \frac{1+|z|}{1-|z|}\right)$$

for $|z| < 1$ where $\Phi(x,y)$ is a universal strictly increasing continuous function for $0 \leq x < \infty$ with $\Phi(0,y) = 0$.

We now prove Theorem 6.1.1.

Proof. Let z_1, z_2, z_3 be distinct points of $\mathbb{D}$ and define

$$h(t) = \frac{f(t,z_1) - f(t,z_2)}{f(t,z_3) - f(t,z_2)}.$$

This is analytic in $\mathbb{D}$ and not equal to 0 or 1, and

$$h(0) = \frac{z_1 - z_2}{z_3 - z_2}.$$

Applying Schottky's theorem gives

$$\left|\frac{f(t,z_1) - f(t,z_2)}{f(t,z_3) - f(t,z_2)}\right| \leq \Phi\left(\left|\frac{z_1 - z_2}{z_3 - z_2}\right|, \frac{1+|t|}{1-|t|}\right). \tag{6.1}$$

Now fix $t \in \mathbb{D}$, z_2 and z_3 and consider z_1 as the variable. Then (6.1) shows that $f(t,z)$ is bounded as z varies in $\mathbb{D}$ for a fixed t. Now let $z_1 \to z_2$, and use the fact that $\Phi(0,y) = 0$ to see that $f(t,z)$ is uniformly continuous in z, and thus has continuous extension to $\overline{\mathbb{D}}$. Hence (6.1) holds for $z_1, z_2, z_3 \in \overline{\mathbb{D}}$, and so $f(t,z)$ is injective for $z \in \overline{\mathbb{D}}$ and therefore a homeomorphism. Now (4.1) and (6.1) show that $f(t,z)$ is quasisymmetric for $z \in \partial\mathbb{D}$ and so $f(t,\partial\mathbb{D})$ is a quasicircle. □

The conclusion of this theorem hints at the following result proved in [34].

Theorem 6.1.3. *A holomorphic motion $f(t,z)$ of $\widehat{\mathbb{C}}$ is quasiconformal in z, and the maximal dilatation of $f(t,.)$ is*

$$\frac{1+|t|}{1-|t|}.$$

If $\mu(t,z)$ is the Beltrami coefficient of $f(t,z)$, then we have $|\mu(t,z)| < 1$ and $\mu(0,z) = 0$,. For almost every $z \in \widehat{\mathbb{C}}$, $\mu(t,z)$ is analytic in t (see [34]), and so the classical Schwarz lemma implies that $|\mu(t,z)| \leq |t|$. This shows that the maximal dilatation of $f(t,z)$ is

$$\frac{1+|t|}{1-|t|}.$$

The study of holomorphic motions have been dominated by the problem of extending a given motion $f : \mathbb{D} \times E \to \widehat{\mathbb{C}}$ to a motion $\widetilde{f} : \mathbb{D} \times F \to \widehat{\mathbb{C}}$, where $E \subset F$.

Definition 6.1.1 can be extended in the obvious way so that we can consider holomorphic motions of E over Ω, where Ω can be any complex manifold. The problem of extending a holomorphic motion $f : \Omega \times E \to \widehat{\mathbb{C}}$, where Ω is specifically a plane domain, can be normed in the following way, and we will see later why this is important. Fix an arbitrary base point $z_0 \in \Omega$. Since the values $f(t, z_0)$ are different for each $t \in E$, we can assume that

$$f(t, z_0) = t \in E.$$

Mark in E two points t_0 and t_1. A linear transformation sends $f(t_0, .)$ and $f(t_1, .)$ to the constant functions 0 and 1 respectively. A holomorphic motion with marked sections $0, 1 \in E$ and $f(0, .) \equiv 0, f(1, .) \equiv 1$ is called normed.

It is well known that the family of all holomorphic maps $\Omega \to \mathbb{C} \setminus \{0, 1\}$ is normal. Hence it is sufficient to solve the problem of extending holomorphic motions under the additional assumptions that E is a finite set, Ω is a domain with compact closure and smooth boundary in $\widehat{\mathbb{C}}$, and all functions $f(t, .)$ are holomorphic in $\widehat{\mathbb{C}}$.

If Ω is not simply connected, then there is a natural toplogical obstruction to the extension of holomorphic motions. Assume $\widetilde{f}(t, .)$ for $t \in \widehat{\mathbb{C}}$ is a holomorphic motion over Ω such that $\widetilde{f}(t, .) = f(t, .)$ for all $t \in E$. If $t \neq t'$, then $\widetilde{f}(t, .) - \widetilde{f}(t', .)$ does not vanish anywhere in Ω, and for any closed curve $\gamma \in \Omega$, the variation of argument

$$\triangle_\gamma \arg(\widetilde{f}(t, .) - \widetilde{f}(t', .)) \in 2\pi\mathbb{Z}$$

is well defined. This number depends continuously on $(t, t') \in \mathbb{C}^2 \setminus \{t = t'\}$, and thus it is constant for all $t \neq t'$. This gives a necessary condition for the existence of a holomorphic extension for a normed motion,

$$\triangle_\gamma \arg(\widetilde{f}(t, .) - \widetilde{f}(t', .)) = 0, \tag{6.2}$$

for $t \neq t'$ in E and for any closed curve $\gamma \in \Omega$.

Exercise 6.1.1. Find a plane domain E and a holomorphic motion of E which cannot be extended to a holomorphic motion of $\widehat{\mathbb{C}}$.

Theorem 6.1.4. *Let Ω be an arbitrary domain in $\widehat{\mathbb{C}}$ and $f(t, .)$ for $t \in E$ be a holomorphic motion of $E \subset \mathbb{C}$ over Ω satisfying (6.2). Then there exists a family of holomorphic functions $\widetilde{f}(t, .)$, with $t \in \widehat{\mathbb{C}}$, defined in Ω with mutually disjoint graphs filling $\Omega \times \widehat{\mathbb{C}}$ such that $\widetilde{f}(t, .) = f(t, .)$ for all $t \in E$. In particular, the holomorphic motion f extends to a holomorphic motion of $\widehat{\mathbb{C}}$.*

This theorem is called Slodkowski's theorem. Important partial results towards this result were obtained in [6] and [53]. The elegant proof we give is due to Chirka in [11].

Proof. We prove the theorem assuming E is finite and $f(t, .)$ are holomorphic in a neighbourhood of $\overline{\Omega}$. We also assume that $z_0 = \infty \in \Omega$.

To demonstrate the method for reducing the problem to the solution of a quasilinear $\overline{\partial}$-equation, we consider the case of a disk, $\Omega = \widehat{\mathbb{C}} \setminus \overline{\mathbb{D}}$. Extend the functions $f(t,.)$, $t \in E$, to the whole of $\widehat{\mathbb{C}}$ by setting

$$f(z) = f\left(\frac{1}{\overline{z}}\right),$$

for $z \in \overline{\mathbb{D}}$. We thus obtain continuous functions with mutually disjoint graphs, that is

$$|f(t,z) - f(t',z)| \geq \delta > 0,$$

for all $z \in \widehat{\mathbb{C}}$, $t \neq t'$. Let $\lambda \in C^\infty(\mathbb{R})$ such that $\lambda(0) = 1$ and $\lambda(t) = 0$ for $t > \delta/2$. The function

$$\phi(z,w) = \sum_{t \in E} \lambda(|w - f(t,z)|) \frac{\partial f(t,z)}{\partial \overline{z}}$$

is uniformly bounded in $\widehat{\mathbb{C}} \times \mathbb{C}$ (first type discontinuities are possible on $\partial\mathbb{D} \times \mathbb{C}$), it vanishes outside $\overline{\mathbb{D}} \times \mathbb{C}$ and satisfies a Lipschitz condition in w with the same constant M for all $z \in \widehat{\mathbb{C}} \setminus \partial\mathbb{D}$. By construction, every $f(t,.)$, $t \in E$, satisfies the quasilinear equation

$$\frac{\partial f}{\partial \overline{z}} = \phi(.,f), \quad f(\infty) = t,$$

where the derivative is in the sense of distributions. For $f \in C(\widehat{\mathbb{C}})$, the equation is equivalent to an integral one,

$$f = t + Kf,$$

where

$$(Kf)(z) = -\frac{1}{\pi} \int_{\mathbb{D}} \frac{\phi(\zeta, f(\zeta))}{\zeta - z} \, \mathrm{d}S_\zeta.$$

The operator K in $C(\widehat{\mathbb{C}})$ has a uniformly bounded image, since ϕ is uniformly bounded, and is compact, because it sends continuous functions to functions of Hölder class with corresponding Hölder constants. The Schauder fixed point theorem implies that for any $t \in \mathbb{C}$, there exists a function $f \in C^\alpha(\widehat{\mathbb{C}})$ such that $f = t + Kf$. This solution is unique, holomorphic in $\widehat{\mathbb{C}} \setminus \overline{\mathbb{D}}$ and we denote it by $\widetilde{f}(t,.)$. Uniqueness, the crucial property of the equation, is a corollary of the Lipschitz condition for ϕ and the following well-known principle: if $h \in C(\widehat{\mathbb{C}})$, $h(a) = 0$, $\partial h/\partial \overline{z} \in L^\infty(\widehat{\mathbb{C}})$ vanishes outside $\overline{\mathbb{D}}$ and

$$\left|\frac{\partial h}{\partial \overline{z}}\right| \leq M|h|$$

pointwise almost everywhere, then h is identically 0. Here, a is an arbitrary point in $\widehat{\mathbb{C}}$, hence the graphs of $\widetilde{f}(t,.)$ are disjoint for different t. Therefore $\widetilde{f}(t,.)$, $t \in \widehat{\mathbb{C}}$, is the desired holomorphic extension.

In the case where Ω is an arbitrary domain with smooth boundary on $\widehat{\mathbb{C}}$, the complement of Ω consists of a finite number of closed domains diffeomorphic to the disk. The extension of $f(t,.)$ to $\widehat{\mathbb{C}}$ described above is not valid in general, but the necessary condition and the following fact help out.

Let the functions $g_0 = 0, g_1, \ldots, g_N \in C^\infty(\partial\mathbb{D})$ have mutually disjoint graphs and

$$\triangle_{\partial\mathbb{D}} \arg(g_j - g_k) = 0$$

for $j \neq k$. Then there exists a diffeomorphism of the cylinder $\overline{\mathbb{D}} \times \mathbb{C}$ of the form

$$(z, w) \mapsto (z, \psi(z, w))$$

such that $\psi(z, g_j(z)) = g_j(1)$ are pairwise different constants. The proof of this fact runs by induction and is omitted.

For a fixed z, define $\psi(z,w) = \psi_z(w)$. By construction, the smooth functions

$$g_j(z) := (\psi_z)^{-1}(g_j(1))$$

have mutually disjoint graphs in $\overline{\mathbb{D}}$, and coincide with the initial ones on $\partial\mathbb{D}$. Applying this procedure to the functions $f(t,.)$, $t \in E$, we obtain a piecewise smooth motion over $\widehat{\mathbb{C}}$ which is holomorphic over Ω.

After that, we construct, as before, the function $\phi(z,w)$, vanishing in $\Omega \times \widehat{\mathbb{C}}$, solve the equation

$$\frac{\partial f}{\partial \bar{z}} = \phi(.,f), \qquad F(\infty) = t$$

in the same way as before and obtain a piecewise smooth foliation of $\widehat{\mathbb{C}} \times \mathbb{C}$ consisting of the graphs of these solutions. Over Ω this gives the sought holomorphic extension of the given holomorphic motion. The proof of the theorem for arbitrary plane domains can be obtained in this way. □

6.2 Equivariant extensions

The following theorem of Earle, Kra, and Krushkal, see [15], is an equivariant version of Slodkowski's theorem.

Theorem 6.2.1. *Let E be a subset of $\widehat{\mathbb{C}}$ that contains at least three points, and let G be a group of Möbius transformations that map E onto itself. Let $f : \mathbb{D} \times E \to \widehat{\mathbb{C}}$ be a holomorphic motion of E. Suppose that for each $g \in G$ and $t \in \mathbb{D}$ there is a Möbius transformation $\theta_t(g)$ such that*

$$f(t, g(z)) = \theta_t(g)(f(t,z)) \tag{6.3}$$

for all $z \in E$. Then f can be extended to a holomorphic motion of $\widehat{\mathbb{C}}$ in such a way that (6.3) holds for all $g \in G$, $t \in \mathbb{D}$ and $z \in \widehat{\mathbb{C}}$.

Proof. We assume that E is closed. In fact, f can be extended to a holomorphic motion of $\overline{E}$, and the extension is a continuous map $\mathbb{D} \times \overline{E} \to \widehat{\mathbb{C}}$. The set $\overline{E}$ is still G-equivariant, and the continuity of f guarantees that (6.3) holds for all $g \in G$, $t \in \mathbb{D}$ and $z \in \overline{E}$. Hence we can assume E is closed.

For each $t \in \mathbb{D}$, $g \in G$, the Möbius transformation $\theta_t(g)$ is completely determined by (6.3), since we assume that E contains at least 3 points. Then for each $t \in \mathbb{D}$, θ_t is a homomorphism of G into the group of Möbius transformations, and θ_0 is the identity. If $g_1, g_2 \in G$ and $\theta_t(g_1) = \theta_t(g_2)$, then (6.3) gives

$$f(t, g_1(z)) = f(t, g_2(z))$$

for all $z \in E$, and therefore $g_1(z) = g_2(z)$ for all $z \in E$. Thus $g_1 = g_2$, and each homomorphism θ_t is injective.

Identifying the complex Lie group $\mathrm{PSL}(2, \mathbb{C})$ with the group of Möbius transformations, for each $g \in G$ we obtain a map $t \mapsto \theta_t(g)$ from $\mathbb{D}$ to $\mathrm{PSL}(2, \mathbb{C})$, and each of these maps are holomorphic. To see this, choose three distinct point $z_1, z_2, z_3 \in E$, and for each $t \in \mathbb{D}$, let h_t be the unique Möbius transformation such that

$$h_t(z_j) = f(t, z_j)$$

for $j = 1, 2, 3$. This combined with (6.3) gives

$$\theta_t(g)(h_t(z_j)) = f(t, g(z_j))$$

for $j = 1, 2, 3$. The right-hand sides of these two equations are holomorphic functions of t for each j, so the maps $t \mapsto h_t$ and $t \mapsto \theta_t(g) \circ h_t$ are holomorphic, and so is $t \mapsto \theta_t(g)$.

The family $\{\theta_t : t \in \mathbb{D}\}$ is a holomorphic family of isomorphisms. Since θ_0 is the identity, θ_t is a quasiconformal deformation for all $t \in \mathbb{D}$. That is, for each $t \in \mathbb{D}$ there is a quasiconformal homeomorphism f_t of $\widehat{\mathbb{C}}$ inducing θ_t, in the sense that

$$f_t \circ g = \theta_t(g) \circ f_t$$

for all $g \in G$. In particular, each of the isomorphisms θ_t is type-preserving.

Let E' be any G-invariant subset of $\widehat{\mathbb{C}}$, which need not be closed, that contains E, and let f be a holomorphic motion of E' extending the given motion E. We say that f is G-equivariant if (6.3) holds for all $t \in \mathbb{D}$, $g \in G$ and $z \in E'$

For each $z \in \widehat{\mathbb{C}}$, let

$$G_z = \{g \in G : g(z) = z\}.$$

The first aim is to extend f to the G-invariant set $E \cup F$, where

$$F = \{z \in \widehat{\mathbb{C}} : G_z \text{ is non-trivial}\}.$$

The set E is closed, G-invariant, and contains at least three points, so it contains the fixed points of all parabolic and loxodromic transformations g in G. To see this consider the orbits of points of E under the action of the cyclic group generated by a parabolic or loxodromic g in G. Thus if $z \in F \setminus E$, the group G_z contains only the identity and elliptic transformations. The same holds for each group $\theta_t(G_z)$, since the isomorphisms θ_t are type-preserving. Therefore all these groups are abelian, and for each $t \in \mathbb{D}$, the non-trivial elements of $\theta_t(G_z)$ all have the same pair of fixed points in $\widehat{\mathbb{C}}$. Since $\theta_t(g)$ depends holomorphically on t for each $g \in G$, then for each $z \in F \setminus E$, there is a unique holomorphic function ϕ_z on $\mathbb{D}$ such that $\phi_z(0) = z$ and $\phi_z(t)$ is fixed by $\theta_t(g)$ for all $g \in G_z$ and $t \in \mathbb{D}$.

Extend f to $E \cup F$ by setting $f(t,z) = \phi_z(t)$ if $z \in \mathbb{D}$ and $z \in F \setminus E$. It is straightforward to check that f is G-equivariant and satisfies properties (1) and (2) of Definition 6.1.1. Proving the injectivity condition of (3) requires more work. To begin with, we claim that if $f(t',z) = f(t',g(z))$ for some $g \in G$, $t \in \mathbb{D}$ and $z \in E \cup F$, then $g \in G_z$. This claim is true if $z \in E$, so to prove it, suppose that $z \in F \setminus E$ and put $w = f(t',z)$ and observe

$$\theta_{t'}(g)(w) = f(t', g(z)) = w$$

because of the G-equivariance of f and the hypothesis. Choose a quasiconformal homeomorphism $f_{t'}$ of $\widehat{\mathbb{C}}$ inducing θ_t and observe that g fixes $z' = (f_{t'})^{-1}(w)$, ie. $g \in G_{z'}$. Similarly, if $h \in G_z$, the G-equivariance implies that $\theta_{t'}(h)$ fixes w and so $h \in G_{z'}$ and $G_z \subset G_{z'}$. Assume for a contradiction that $g \notin G_z$, so $z' \neq z$, and choose a non-trivial $h \in G_z$. The commutator $h^* = h \circ g \circ h^{-1} \circ g^{-1}$ is parabolic, so z' is its only fixed point. The transformation $\theta_{t'}(h^*)$ is also parabolic and fixes both $f(t',z')$ (by the G-equivariance) and w (because $f_{t'}$ induces $\theta_{t'}$). Therefore $f(t',z') = w$. Since $h \in G_z$ and $G_z \subset G_{z'}$, G-equivariance implies that $\theta_t(h)$ fixes both $f(t,z)$ and $f(t,z')$ for every $t \in \mathbb{D}$. However, $\theta_t(h)$ is always elliptic and its fixed points are given by two holomorphic functions of t in $\mathbb{D}$ with disjoint graphs. Since $f(0,z) \neq f(0,z')$, then $f(t,z) = f(t,z')$ for all $t \in \mathbb{D}$, contradicting the equality when $t = t'$, which proves the claim above.

Now we can prove the injectivity property of the extension f to $E \cup F$. Suppose $f(t',z) = f(t',z')$ for some $t' \in \mathbb{D}$ and $z, z' \in E \cup F$. The aim is to show that $z = z'$. This is true if both z and z' are in E, so assume that $z \in F \setminus E$. For all $g \in G$,

$$f(t', g(z)) = \theta_{t'}(g)(f(t',z)) = \theta_{t'}(g)(f(t',z')) = f(t', g(z'))$$

so $f(t',z) = f(t',g(z))$ if $g \in G_{z'}$. From the claim above, $G_{z'} \subset G_z$, and so by symmetry, $G_{z'} = G_z$. Since $z \in F \setminus E$, G_z is a non-trivial abelian group. If $z \neq z'$, then z and z' are the two fixed points of G_z, so $f(t,z)$ and $f(t,z')$ are the two fixed points of $\theta_t(g)$ for any $t \in \mathbb{D}$ and non-trivial $g \in G$. This contradicts the assumption that $f(t',z) = f(t',z')$ and so $z = z'$ and we have the injectivity property. Thus we have extended f to a G-equivariant holomorphic motion of $E \cup F$.

Now let E' be any G-invariant subset $\widehat{\mathbb{C}}$ such that $E \cup F \subset E'$ and let $z_0 \in \widehat{\mathbb{C}} \setminus E'$, and put

$$E'' = E' \cup \{g(z_0) : g \in G\}.$$

We claim that every G-equivariant holomorphic motion of E' extends to a G-equivariant holomorphic motion of E''. To verify the claim, let $f : \mathbb{D} \times E' \to \widehat{\mathbb{C}}$ be any G-equivariant holomorphic motion of E'. Extend it (by the extension theorem of Slodkowski) to a holomorphic motion of $E' \cup \{z_0\}$ that we still denote by f. To extend this motion to E'', set

$$f(t, g(z_0)) = \theta_t(g)(f(t, z_0))$$

for $t \in \mathbb{D}$ and $g \in G$. We have to show that the extension is a well-defined G-equivariant holomorphic motion of E''. If $g_1(z_0) = g_2(z_0)$ for $g_1, g_2 \in G$, then $g_1 = g_2$ since $z_0 \notin F$, and this f is well defined. The G-equivariance follows from the definition of f, and properties (1) and (2) of a holomorphic motion are easily verified. We return to property (3) after discussing the fixed points of transformations in $\theta_t(G)$ for a given $t \in \mathbb{D}$.

Let F_g be the set of fixed points of a non-trivial $g \in G$. For any subset X of E', define

$$f(t, X) = \{\zeta \in \widehat{\mathbb{C}} : \zeta = f(t, z) \text{ for some } z \in X\}.$$

For any non-trivial $g \in G$, the set $f(t, F_g)$ is contiained in the fixed point set of $\theta_t(g)$ because f is G-equivariant. In fact, $f(t, F_g)$ is precisely the fixed point set of $\theta_t(g)$, since θ_t is type-preserving. Thus all fixed points of non-trivial elements of $\theta_t(G)$ belong to the set $f(t, F)$, which by hypothesis is contained in $f(t, E')$.

To verify (3), assume that $f(t, z_1) = f(t, z_2)$ for some $t \in \mathbb{D}$ and also that $z_1, z_2 \in E''$, and we must show that $z_1 = z_2$. This is true by construction if z_1 and z_2 both belong to $E' \cup \{z_0\}$, so assume that $z_1 = g(z_0)$ for some non-trivial $g \in G$. By G-equivariance, $f(t, z_0) = f(t, g^{-1}(z_2))$. If $z_2 \in E'$, then we have that $f(t, g^{-1}(z_2)) \in f(t, E')$. Since $f(t, z_0) \notin f(t, E')$, then $z_2 = h(z_0)$ for some $h \in G$. Then, by G-equivariance, $f(t, z_0)$ is a fixed point of

$$\theta_t(g)^{-1} \circ \theta_t(h) = \theta_t(g^{-1} \circ h).$$

Since $f(t, z_0) \notin f(t, E')$, then $g = h$ and $z_1 = z_2$ as required. This shows that f can be extended to a G-equivariant holomorphic motion of E''.

To prove the theorem, iterate the construction above, applying it first to $E' = E \cup F$, and choosing successive points z_0 so that the union of their G-orbits is dense in $\widehat{\mathbb{C}}$. The sequence of extended motions converges to the desired G-equivariant holomorphic motion of $\widehat{\mathbb{C}}$. □

6.3 Area distortion

In this section, we will illustrate an application of holomorphic motions by proving the following theorem of Astala, [5], partially proved in the earlier paper [24]. The proof below is due to Eremenko and Hamilton [19].

Theorem 6.3.1. *Let f be a K-quasiconformal mapping of $\mathbb{D}$ onto itself satisfying $f(0) = 0$. Then for any measurable $E \subset \mathbb{D}$,*

$$|f(E)| \leq C(K)|E|^{1/K} \tag{6.4}$$

where $C(K)$ is a constant depending only on K, and $|.|$ denotes area.

See Exercise 3.1.1 for an example of a mapping which achieves equality in (6.4). We first need a lemma from statistical mechanics.

Lemma 6.3.1. *Let $p_j > 0$ and $q_j > 0$ be probability distributions on the set $\{1, \ldots, n\}$. Then*

$$-\sum_{j=1}^{n} p_j \log q_j + \sum_{j=1}^{n} p_j \log p_j \geq 0.$$

Proof. Let $\varphi(x) = x \log x$. Then the left-hand side of the inequality is

$$\sum q_j \varphi\left(\frac{p_j}{q_j}\right).$$

Since φ is convex,

$$\sum q_j \varphi\left(\frac{p_j}{q_j}\right) \geq \varphi\left(\sum q_j \frac{p_j}{q_j}\right) = \varphi(1) = 0.$$

□

Lemma 6.3.2. *Let $a_1, \ldots, a_n$ be positive functions on $\mathbb{D}$ such that $\log a_j$ is harmonic for every j and*

$$\sum_{j=1}^{n} a_j(\lambda) \leq 1$$

for $|\lambda| < 1$. Then, for $|\lambda| < 1$,

$$\sum_{j=1}^{n} a_j(\lambda) \leq \left(\sum_{j=1}^{n} a_j(0)\right)^{(1-|\lambda|)/(1+|\lambda|)}.$$

Proof. For $|\lambda| < 1$ and $|z| < 1$ define the probability distributions

$$p_j = \frac{a_j(\lambda)}{\sum a_j(\lambda)},$$
$$q_j = \frac{a_j(z)}{\sum a_j(z)}.$$

Fix λ, and set

$$H(z) = -\sum p_j \log a_j(z) + \sum p_j \log p_j.$$

Observe that H is harmonic in z, and by Lemma 6.3.1, $H(z) \geq -\log \sum a_j(z) \geq 0$. By Harnack's inequality,

$$H(z) \geq \left(\frac{1-|z|}{1+|z|}\right) H(0).$$

Put $z = \lambda$, and use Lemma 6.3.1 to obtain

$$H(\lambda) = -\log \sum a_j(\lambda) \geq \left(\frac{1-|\lambda|}{1+|\lambda|}\right)\left(-\sum p_j \log a_j(0) + \sum p_j \log p_j\right)$$
$$\geq \left(\frac{1-|\lambda|}{1+|\lambda|}\right)\left(-\log \sum a_j(0)\right),$$

which completes the proof. $\square$

We will use a continuous version of Lemma 6.3.2. If $E \subset \mathbb{D}$, $a(z,\lambda)$ is defined on $E \times \mathbb{D}$, $\log a(z,\lambda)$ is harmonic in λ and

$$\int_E a(z,\lambda)\, dx\, dy \leq 1$$

where $z = x + iy$ and $|\lambda| < 1$, then

$$\int_E a(z,\lambda)\, dx\, dy \leq \left(\int_E a(z,0)\, dx\, dy\right)^{(1-|\lambda|)/(1+|\lambda|)}.$$

Theorem 6.3.2. *Let f be a K-quasiconformal mapping of $\mathbb{C}$ which is conformal on $\mathbb{C} \setminus \overline{\mathbb{D}}$ and $f(z) = z + o(1)$ as $z \to \infty$.*

(a) *If f is conformal on $E \subset \overline{\mathbb{D}}$, then*

$$|f(E)| \leq \pi^{1-1/K}|E|^{1/K}.$$

(b) *If $E \subset \overline{\mathbb{D}}$ with f conformal on $\mathbb{C} \setminus E$, then*

$$|f(E)| \leq K|E|.$$

(c) *Thus in general for $E \subset \overline{\mathbb{D}}$,*

$$|f(E)| \leq K\pi^{1-1/K}|E|^{1/K}.$$

Proof. Suppose f has complex dilatation μ supported on $\overline{\mathbb{D}}$ and, without loss of generality, assume that μ is smooth (since the smooth case is dense, a uniform bound for the smooth case yields the general case), and define f_λ, a K_λ-quasiconformal map of $\mathbb{C}$ which is conformal on $\mathbb{C} \setminus \mathbb{D}$ and $f_\lambda(z) = z + o(1)$ as $z \to \infty$, where $K_\lambda = (1+|\lambda|)/(1-|\lambda|)$ and f_λ has dilatation

$$\mu_\lambda(z) = \lambda \frac{K+1}{K-1}\mu(z)$$

where $|\lambda| < 1$. Now f_λ has Jacobian

$$J_\lambda(z) = |(f_\lambda)_z|^2(1 - |\mu_\lambda|^2)$$

and since the dilatations are smooth, this is non-zero everywhere. If f is conformal on E, define

$$a(z,\lambda) = \frac{1}{\pi}|(f_\lambda)_z|^2.$$

From the remark following Corollary 5.3.1, $(f_\lambda)_z$ is holomorphic in λ and so $\log a(z,\lambda)$ is harmonic for $|\lambda| < 1$ and $z \in E$. Since $f(z) = z + o(1)$ as $z \to \infty$, the classical Area theorem for conformal mappings gives

$$\int_{\mathbb{D}} J_\lambda(z)\,\mathrm{d}x\,\mathrm{d}y \le \pi.$$

Therefore $a(z,\lambda)$ satisfies the continuous version of Lemma 6.3.2, and so

$$\frac{1}{\pi}\int_E J_\lambda(z) \le \left(\frac{|E|}{\pi}\right)^{(1-|\lambda|)/(1+|\lambda|)},$$

and noting that $|\mathbb{D}| = \pi$. Setting $\lambda = (K-1)(K+1)$ gives $\mu_\lambda = \mu$ and

$$|f(E)| \le \pi^{1-1/K}|E|^{1/K}$$

which proves (a). Moving onto (b), first recall that the Hilbert transform T defined in (5.4) is a unitary transformation of $L^2(\mathbb{C})$ and so for any set G

$$\int_G |T(\chi_G)|\,\mathrm{d}x\,\mathrm{d}y \le |G|,$$

where χ_G is the indicator function for G. Therefore for any function ρ supported on G,

$$\left|\int_G T(\rho)\,\mathrm{d}x\,\mathrm{d}y\right| \le ||\rho||_\infty |G|.$$

Now, for any function μ with $||\mu||_\infty = 1$ supported on E, define $\mu_t(z) = t\mu(z)$ and the corresponding family of normalized maps f_t for $0 < t < 1$, where $f_0(z) = z$ and $f_{|\lambda|} = f$. This can be realized as a deformation family of quasiconformal maps

$$\frac{\partial f_t}{\partial t} = g_t \circ f_t,$$

$$g_t(z) = z + P\rho_t,$$

$$\rho_t = \frac{\mu \circ f_t^{-1}}{1 - t^2|\mu \circ f_t^{-1}|^2}\mathrm{e}^{2i\arg((f_t^{-1})_z}$$

by the composition formula for complex dilatations, and where P is the Cauchy transform. Since $\partial S = T$,

$$\frac{\mathrm{d}|f_t(E)|}{\mathrm{d}t} = 2\Re\int_{f_t(E)} T(\rho_t)\,\mathrm{d}x\,\mathrm{d}y$$

and so from the inequality for T above

$$\frac{\mathrm{d}|f_t(E)|}{\mathrm{d}t} \le \frac{2|f_t(E)|}{1-t^2}.$$

Integrating this gives

$$|f_t(E)| \le \frac{1+t}{1-t}|E|,$$

which proves (b). For (c), write $f = g \circ h$ where h is conformal on E, and g is conformal on $\mathbb{C} \setminus h(E)$. Then h has dilatation $\mu(z)$ on $\overline{\mathbb{D}} \setminus E$ and zero elsewhere, and g has dilatation $\mu(h^{-1}(z))$ on $h(E)$ and zero elsewhere. Note that h is normalized and so is g, as $h(\overline{\mathbb{D}})$ has transfinite diameter 1. $\square$

Remark. For any function μ with $||\mu||_\infty < 1$ supported on $\overline{\mathbb{D}} \setminus E$, define the functions $\mu_\lambda(z) = \lambda\mu(z)$ and the corresponding family of normalized maps f_λ. Letting $\lambda \to 0$ shows that

$$\begin{aligned} |f_\lambda(E)| &= |E| + 2\Re\left(\lambda \int_E T(\mu)\, dx\, dy\right) + o(\lambda) \\ &\leq \pi^{2\lambda+o(\lambda)}|E|^{1-2\lambda+o(\lambda)} = |E| + 2|\lambda||E| \log\frac{\pi}{|E|} + o(\lambda) \end{aligned}$$

by part (a) of Theorem 6.3.2. Hence

$$\left|\int_E T(\mu)\, dx\, dy\right| \leq |E| \log\frac{\pi}{|E|},$$

and so, as in the proof of (b) in Theorem 6.3.2, for all μ supported on $\overline{\mathbb{D}} \setminus E$ and $||\mu||_\infty < 1$,

$$\left|\int_{\overline{\mathbb{D}}\setminus E} T(\chi_E)\overline{\mu}(z)\, dx\, dy\right| \leq |E| \log\frac{\pi}{|E|}.$$

7

Teichmüller spaces

This chapter introduces the idea of Teichmüller spaces as equivalence classes of quasiconformal maps. We show how this is defined for the universal Teichmüller space $T(\mathbb{H})$, and then how this can be generalized for Teichmüller spaces of Riemann surfaces. We define the Teichmüller metric on Teichmüller space. As an extended example, the Teichmüller space of the torus is computed. We show how the Bers embedding can embed Teichmüller space $T(M)$ into $Q(M)$, the Banach space of quadratic differentials on M, and how to use this embedding to view $T(M)$ as a complex Banach manifold modelled on $Q(M)$.

7.1 Universal Teichmüller space

Let $\mathcal{F}$ be the family of all quasiconformal self-mappings of $\mathbb{H}$ which are normalized so that their extensions to $\mathbb{R}$ fix $0, 1$, and ∞. From the solution of the Beltrami differential equation in Theorem 5.2.2, there is a one-to-one correspondence between $\mathcal{F}$ and the open unit ball B of $L^\infty(\mathbb{H})$. Now, we can put an equivalence relation $\sim$ on elements of $\mathcal{F}$ by declaring that $f_1 \sim f_2$ if and only if the extensions of f_1 and f_2 to $\mathbb{R}$ agree on $\mathbb{R}$. Equivalently, two elements μ_1 and μ_2 of B are related by $\sim$ if and only if the extensions of the normal solutions f^{μ_1} and f^{μ_2} to $\mathbb{R}$ agree on $\mathbb{R}$.

Definition 7.1.1. *The set of equivalence classes of $\mathcal{F}$ under $\sim$ is called the universal Teichmüller space $T(\mathbb{H})$.*

We can also consider the universal Teichmüller space as the set of equivalence classes of B with respect to $\sim$. Let QS be the space of quasisymmetric functions on $\mathbb{R}$ which are normalized to fix 0, 1 and ∞. For $f \in \mathcal{F}$, let $[f]$ be the corresponding element of $T(\mathbb{H})$. Then, from Theorem 4.2.1, the mapping

$$[f] \to f|_{\mathbb{R}},$$

where we consider the extension of f to $\mathbb{R}$ is a bijection between $T(\mathbb{H})$ and QS. Therefore $T(\mathbb{H})$ can be considered as the set of normalized quasisymmetric functions. The obvious group operation $[f] \circ [g] = [f \circ g]$ makes $T(\mathbb{H})$ into a group, which the reader can check. The space of normalized quasisymmetric functions, QS, is also a group under composition, and then the mapping $T(\mathbb{H}) \to$ QS is a group isomorphism.

Now let $f \in \mathcal{F}$ and $g : \mathbb{H} \to \mathbb{H}$ be quasiconformal. Normalize with an appropriate Möbius transformation h so that $h \circ f \circ g^{-1} \in \mathcal{F}$. Define the mapping $w = w_{[g]} : T(\mathbb{H}) \to T(\mathbb{H})$ given by

$$w([f]) = [h \circ f \circ g^{-1}].$$

The mapping w is a well-defined bijection with $w_{[0]}$ the identity mapping and

$$w_{[g_1]} \circ w_{[g_2]} = w_{[g_1] \circ [g_2]}.$$

This makes the collection $\{w_{[g]} : g \text{ quasiconformal}\}$ into a group M, which is called the universal modular group. In the special case $g \in \mathcal{F}$, $w_{[g]}([f]) = [f \circ g^{-1}]$ and the corresponding subgroup M_t of M is the group of right translations of $T(\mathbb{H})$. If $[f_1], [f_2] \in T(\mathbb{H})$ and $g = f_2^{-1} \circ f_1$, then $[f_2] = w_{[g]}([f_2])$, which shows that M_t is transitive.

Recall that f^μ is the normalized quasiconformal self-mapping of $\mathbb{H}$ with Beltrami coefficient $\mu \in B$. Now define f_μ to be the normalized quasiconformal mapping of $\mathbb{C}$ which has Beltrami coefficient equal to μ on $\mathbb{H}$ and 0 on $\mathbb{R} \cup \mathbb{H}^-$. By Theorem 2.3.2, $f_\mu|_{\mathbb{H}^-}$ is conformal.

Theorem 7.1.1. *Given $\mu, \nu \in B$ and the equivalence relation $\sim$ defined above,*

$$\mu \sim \nu \Leftrightarrow f_\mu|_{\mathbb{H}^-} = f_\nu|_{\mathbb{H}^-}.$$

Proof. First assume that $f_\mu|_{\mathbb{H}^-} = f_\nu|_{\mathbb{H}^-}$. Both of the mappings $f_\mu \circ (f^\mu)^{-1}$ and $f_\nu \circ (f^\nu)^{-1}$ are conformal on $\mathbb{H}$, both map $\mathbb{H}$ onto the same quasidisk, and fix $0, 1$, and ∞. Hence they agree on $\mathbb{H}$ and $\mathbb{R}$. Since f_μ and f_ν agree on $\mathbb{R}$, then f^μ and f^ν agree on $\mathbb{R}$, or equivalently $\mu \sim \nu$.

Now assume that f^μ and f^ν agree on $\mathbb{R}$. Define $g : \mathbb{C} \to \mathbb{C}$ by $g = f_\mu \circ f_\nu^{-1}$ on $f_\nu(\mathbb{R} \cup \mathbb{H}^-)$ and $g = f_\mu \circ (f^\mu)^{-1} \circ f^\nu \circ f_\nu^{-1}$ on $f_\nu(\mathbb{H})$. Since f^μ and f^ν agree on $\mathbb{R}$, g is a homeomorpism of $\mathbb{C}$, and $g|_{f_\nu(\mathbb{H}^-)}$ is conformal. Since $f_\mu \circ (f^\mu)^{-1}$ and $f^\nu \circ f_\nu^{-1}$ are both conformal, then $g|_{f_\nu(\mathbb{H})}$ is also conformal. This means that g is a Möbius transformation, and because of the normalization, g must be the identity and so f_μ and f_ν are equal on $\mathbb{H}^-$. □

Using this theorem gives another model of the universal Teichmüller space, that of the set of normalized conformal mappings $f_\mu|_{\mathbb{H}^-}$.

7.2 Teichmüller space of a Riemann surface

We first introduce the automorphism group of a Riemann surface. If M and N are Riemann surfaces, and $f : M \to N$ is conformal, then we say that M

is conformally equivalent to N. If f is quasiconformal, then M and N are quasiconformally equivalent. The automorphism group of M is

$$\mathrm{Aut}(M) = \{f : M \to M : f \text{ is conformal}\}.$$

For example, $\mathrm{Aut}(\mathbb{D})$ is the group of Möbius transformations from $\mathbb{D}$ to $\mathbb{D}$. Recall from Exercise 1.5.1 that $\mathrm{Aut}(M)$ is finite when M is a closed Riemann surface, thus if $f : M \to M$ is conformal, then there exists $n \in \mathbb{N}$ such that f^n is the identity.

If $f : M \to N$ is quasiconformal, then by definition M and N are homeomorphic, but not all homeomorphic surfaces are quasiconformally equivalent. For example, $M = \mathbb{D} \setminus \{0\}$ and $N = \mathbb{D} \setminus \{|z| \leq 1/2\}$ are not quasiconformally related, but they are homeomorpic. The following theorem shows that more can be said in the case of closed surfaces. For more on quasiconformal isotopies, see [16].

Theorem 7.2.1. *If $f : M \to N$ is a homeomorphism between closed surfaces, then f is homotopic to a quasiconformal map.*

Proof. Since M and N are homeomorphic closed surfaces, they both have the same genus and no boundary. Now, since M is compact, it has a finite covering by domains $U_1, \ldots, U_n$, each of which is conformally equivalent to $\mathbb{D}$ and such that ∂U_i is an analytic curve for each i. Define $f_0 = f$ and inductively define the mappings f_i for $i = 1, \ldots, n$ in the following way: $f_i = f_{i-1}$ on $M \setminus U_i$, and inside U_i, f_i is the Beurling–Ahlfors extension (see the proof of Theorem 4.2.1) of the boundary values $f_{i-1}|_{\partial\partial U_i}$. Since M is compact, f_n is a quasiconformal mapping of M. Now the mapping given by

$$(p, t) \mapsto t f_k(p) + (1 - t) f_{k-1}(p),$$

where the '+' implicitly assumes a lift of $f_k(p)$ and $f_{k-1}(p)$ to a chart in $\mathbb{C}$, defines a homotopy between f_{k-1} and f_k, and hence f_n, which is quasiconformal, is homotopic to f, which proves the theorem. □

The deformation space $\mathrm{Def}(M)$ of a Riemann surface M is the set of pairs (N, f) where N is a Riemann surface, and $f : M \to N$ is a quasiconformal map. For any plane domain Ω with its appropriate Riemann map f, we have $(\Omega, f) \in \mathrm{Def}(\mathbb{D})$. An equivalence relation $\sim$ can be defined on $\mathrm{Def}(M)$ by requiring that $(N_1, f_1) \sim (N_2, f_2)$ if and only if

$$f_1 \circ f_2^{-1} : N_2 \to N_1$$

is homotopic to a conformal map $g : N_2 \to N_1$.

Exercise 7.2.1. Show that two maps f and g between hyperbolic Riemann surfaces are homotopic if they can be lifted to mappings of $\mathbb{H}$ which agree on $\partial\mathbb{R}$.

Definition 7.2.1. *The Teichmüller space $T(M)$ of a Riemann surface M is given by*

$$\frac{\mathrm{Def}(M)}{\sim}.$$

This definition agrees with the definition of universal Teichmüller space given in Section 7.1. To see this, first note that since all the quasiconformal images of $\mathbb{H}$ are conformally equivalent, only the normalized quasiconformal self-mappings f^{μ} of $\mathbb{H}$ need be considered. Then $f^{\mu_2} \circ (f^{\mu_1})^{-1}$ is homotopic to a conformal map if and only if $f^{\mu_2} \circ (f^{\mu_1})^{-1}$ is the identity mapping on $\mathbb{R}$. Therefore f^{μ_1} agrees with f^{μ_2} on $\mathbb{R}$ which is the definition for f^{μ_1} and f^{μ_2} to determine the same point of universal Teichmüller space.

As in the considerations for the universal Teichmüller space, in view of Theorem 5.2.4, $T(M)$ can be considered as a space of Beltrami differentials under the corresponding equivalence relation. Given a Beltrami differential μ on a Riemann surface M, it lifts to a Beltrami differential $\widetilde{\mu}$ on $\mathbb{H}$ which satisfies $\widetilde{\mu} = (\widetilde{\mu} \circ g)\overline{g'}/g'$ for every g in the covering group of M. We write f_{μ} for the quasiconformal mapping of the plane which has the complex dilatation $\widetilde{\mu}$ on $\mathbb{H}$, and 0 on $\mathbb{H}^{-}$.

Theorem 7.2.2. *The Beltrami differentials μ and ν on M are equivalent under $\sim$ if and only if $f^{\mu}|_{\mathbb{R}} = f^{\nu}|_{\mathbb{R}}$ or if and only if $f_{\mu}|_{\mathbb{H}^-} = f_{\nu}|_{\mathbb{H}^-}$.*

Proof. First assume that μ and ν are equivalent. Let g and h be quasiconformal mappings of M which lift to f^{μ} and f^{ν} respectively. Then there is a conformal map $\alpha : g(M) \to h(M)$ such that $\alpha \circ g$ is homotopic to h. Now $f^{\nu} = A \circ f^{\mu}$ on $\mathbb{R}$, where A is a Möbius transformation, since it is a lift of α. Since f^{μ} and f^{ν} both fix $0, 1$, and ∞, A must be the identity.

For the converse, suppose $f^{\mu} = f^{\nu}$ on $\mathbb{R}$. Then f^{μ} and f^{ν} induce the same isomorphism of the covering group $\mathbb{H}$ over M onto a Fuchsian group Γ. The projections of f^{μ} and f^{ν} map M onto the same Riemann surface $\mathbb{H}/\Gamma$, and these projections are homotopic. This means that μ and ν are equivalent.

Using the method of Theorem 7.1.1 proves the rest of this theorem. □

There is a space analogous to Teichmüller space, but defined using the notion of asymptotically conformal mappings.

Definition 7.2.2. *A map $f : M \to N$ between Riemann surfaces is called asymptotically conformal if for all $\epsilon > 0$, there exists a compact subset E of M such that the dilatation $D_f(z) < 1 + \epsilon$ for all $z \in M \setminus E$.*

As above, we can put an equivalence relation $\sim_{AT}$ on $\mathrm{Def}(M)$ by requiring that $(N_1, f_1) \sim_{AT} (N_2, f_2)$ if and only if

$$f_1 \circ f_2^{-1} : N_2 \to N_1$$

is homotopic to an asymptotically conformal map $g : N_2 \to N_1$.

Definition 7.2.3. *Asymptotic Teichmüller space* AT(M) *is defined to be*

$$\frac{\mathrm{Def}(M)}{\sim_{\mathrm{AT}}}.$$

Exercise 7.2.2. If M is a compact Riemann surface, what can be said about AT(M)?

7.3 Teichmüller metric

For $[f_0], [g_0] \in T(M)$, we can define the Teichmüller distance

$$d_T([f_0],[g_0]) = \frac{1}{2} \inf \log K_{g \circ f^{-1}} \tag{7.1}$$

where the infimum is taken over all maps f and g in the Teichmüller classes of f_0 and g_0 respectively, and recalling that K_f is the maximal dilatation of f. We write d_T instead of $d_{T(M)}$ for brevity where it is clear which Teichmüller space $[f]$ and $[g]$ are in. Since $K_f = K_{f^{-1}}$, then

$$d_T([f],[g]) = d_T([g],[f]).$$

Using Theorem 1.4.1,

$$K_{g \circ f^{-1}} K_{f \circ h^{-1}} \geq K_{g \circ h^{-1}}$$

which shows that

$$d_T([f],[g]) \leq d_T([f],[g]) + d_T([g],[h]).$$

So the only requirement left for d_T to be a metric is that $d_T([f],[g]) = 0$ implies $[f] = [g]$. To show this, we need the following result.

Theorem 7.3.1. *Let $f : M \to N$ be quasiconformal. Then the set of all quasiconformal mappings of M onto N which are homotopic to f contains an extremal mapping with the smallest maximal dilatation.*

Proof. If the universal covering surface is $\widehat{\mathbb{C}}$ or $\mathbb{C}$ and M is not compact, then the theorem is trivial. We omit the case where M is a torus, the proof for which follows the same lines as below. This leaves the case where $\mathbb{H}$ is the universal cover of M. Lift each quasiconformal mapping g, which is homotopic to f, to a mapping $\widetilde{g} : \mathbb{H} \to \mathbb{H}$ such that all mappings $\widetilde{g}$ agree on $\mathbb{R}$. The set of these lifts contains its quasiconformal limits, and so there exists a mapping $\widetilde{g_0}$ with smallest maximal dilatation. The projection g_0 is the required extremal mapping. □

Now given $[f], [g] \in T(M)$, fix $f_0 \in [f]$ and $g_0 \in [g]$, and let F be the class of all quasiconformal mappings of $f_0(M)$ onto $g_0(M)$ homotopic to $g_0 \circ f_0^{-1}$. We see that

$$d_T([f],[g]) = \frac{1}{2} \inf_{h \in F} \log K_h.$$

Theorem 7.3.1 shows that inf can be replaced by min, and so if $d_T([f],[g]) = 0$, then the class F contains a conformal mapping, and so $[f] = [g]$. This shows that $(T(M), d_T)$ is a metric space. The Teichmüller metric can be expressed in terms of Beltrami differentials,

$$d_T([\mu_0],[\nu_0]) = \frac{1}{2} \inf_{\mu\in[\mu_0],\nu\in[\nu_0]} \log \frac{1 + ||(\mu-\nu)/(1-\overline{\mu}\nu)||_\infty}{1 - ||(\mu-\nu)/(1-\overline{\mu}\nu)||_\infty}.$$

A Beltrami differential $\mu \in [\mu_0]$ is called extremal if $||\mu||_\infty \le ||\nu||_\infty$ for any $\nu \in [\mu_0]$.

Theorem 7.3.2. *Let $[\mu_0] \in T(M)$. If μ is an extremal Beltrami differential in the Teichmüller class $[\mu_0]$, then*

$$\mu_t = \frac{(1+|\mu|)^t - (1-|\mu|)^t}{(1+|\mu|)^t + (1-|\mu|)^t} \cdot \frac{\mu}{|\mu|} \tag{7.2}$$

for $0 \le t \le 1$ is extremal for $[\mu_t] \in T(M)$. The arc $t \to [\mu_t]$ is a geodesic from 0 to $[\mu_t]$, and $d_T([\mu_t],0) = t d_T([\mu],0)$.

Proof. Since μ represents a point of $T(M)$, it is a Beltrami differential for the covering group G of the universal covering surface over M, and so $\mu = (\mu \circ g)\overline{g'}/g'$ for every $g \in G$. By inspection, μ_t defined in (7.2) also satisfies this condition, and thus $[\mu_t]$ is indeed a point of $T(M)$.

Recalling Theorem 5.2.5, if f^μ has maximal dilatation K, then f^{μ_t} has maximal dilatation K^t, and $f^\mu \circ (f^{\mu_t})^{-1}$ has maximal dilatation K^{1-t}. Now if $g \in [\mu_t]$, then

$$\widetilde{g} = f^\mu \circ (f^{\mu_t})^{-1} \circ g \in [\mu_t]$$

and so $K \le K_{\widetilde{g}} \le K^{1-t} K_g$, which means that $K_g \ge K^t$. Thus μ_t is extremal for $[\mu_t]$. Using this method, we see that

$$d_T([\mu_t],0) = \frac{1}{2}\log K^t = t d_T([\mu],0).$$

Since $f^\mu \circ (f^{\mu_t})^{-1}$ has maximal dilatation K^{1-t}, then

$$d_T([\mu_t,[\mu]) \le (1-t) d_T([\mu],0).$$

Thus for every t,

$$d_T(0,[\mu_t]) + d_T([\mu_t],[\mu]) = d_T(0,[\mu])$$

and repeating this argument for an arbitrary subarc $t \to [\mu_t]$ shows that it is a geodesic. We see in later chapters examples where μ is not uniquely extremal, and therefore the geodesic need not be unique. □

This theorem gives the following immediate corollary.

Theorem 7.3.3. *The Teichmüller space $T(M)$ is pathwise connected.*

Proof. The geodesic $t \to [\mu_t]$ is a path joining the base-point in $T(M)$ to the point $[\mu] \in T(M)$, as is the path $t \to [t\mu]$, which completes the proof. □

We now prove another topological property of Teichmüller space, namely that $(T(M), d_T)$ is complete.

Lemma 7.3.1. *Every Cauchy sequence in $(T(M), d_T)$ contains a subsequence whose points are represented by Beltrami differential μ_n such that*

(a) $||\mu_n - \mu||_\infty \to 0$,
(b) $\lim \mu_n(z) = \mu(z)$ *exists almost everywhere,*
(c) $[f_{\mu_n}] \to [f_\mu]$ *in the Teichmüller metric,*
(d) $f_{\mu_n}(z) \to f_\mu(z)$ *uniformly in the spherical metric,*
(e) $f^{\mu_n}(z) \to f^\mu(z)$ *uniformly on compact subsets of* $\mathbb{H}$.

Proof. If G is the covering group for $\mathbb{H}$ over M, then each μ_n is a Beltrami differential for G, that is, for each n and $g \in G$, $\mu_n = (\mu_n \circ g)\overline{g'}/g'$. Thus once we have shown property (b), μ is also a Beltrami differential for G.

To simplify notation, we will write f_n for f_{μ_n} and renumber the functions every time we pass to a subsequence. Now let $([f_n])$ be a Cauchy sequence in $(T(M), d_T)$. We will construct a subsequence (f_n) inductively which satisfies the conclusions of the lemma. First fix a mapping f_i such that for $j = 1, 2, \ldots,$

$$\inf \log K_{f_{i+j} \circ f_i^{-1}} < \frac{1}{2}$$

where the infimum is taken over all mappings in $[f_{i+j}]$ for each j. The existence of such a mapping f_i is guaranteed since $([f_n])$ is a Cauchy sequence. Renumerate the sequence by setting $f_i = f_1$. Now choose f_n from its equivalence class such that

$$\log K_{f_n \circ f_1^{-1}} < \frac{1}{2}.$$

From this new sequence (f_n), choose f_k so that

$$\inf \log K_{f_{k+j} \circ f_k^{-1}} < \frac{1}{4}$$

where for each j the infimum is taken over all mappings in the class $[f_{k+j}]$. Set $f_k = f_2$ and for $n > 2$ choose a representative of $[f_n]$ such that

$$\log K_{f_n \circ f_2^{-1}} < \frac{1}{4}.$$

Continuing this procedure gives a sequence (f_n) such that $([f_n])$ is a subsequence of the given Cauchy sequence and for any two consecutive indices, the maximal dilatations satisfy

$$\log K_{f_{n+1} \circ f_n^{-1}} < 2^{-n}$$

for $n = 1, 2, \ldots$. Therefore,

$$\log K_{f_{n+j} \circ f_n^{-1}} \leq \sum_{i=1}^{j} 2^{-(n+i-1)} < 2^{-n+1} \tag{7.3}$$

for $n, j = 1, 2, \ldots$. The complex dilatations μ_n of f_n satisfy

$$||\mu_{n+j} - \mu_n||_\infty \leq 2 \left|\left| \frac{\mu_{n+j} - \mu_n}{1 - \overline{\mu_n}\mu_{n+j}} \right|\right|_\infty < 2\tanh 2^{-n}.$$

We conclude that (μ_n) is a Cauchy sequence in L^∞. Since L^∞ is complete, the limit μ of μ_n as $n \to \infty$ exists in L^∞ and

$$||\mu_n - \mu||_\infty \to 0$$

as $n \to \infty$. This proves (a) and (b). Using (7.3), the mappings f_n, and also f^{μ_n}, are K-quasiconformal for a fixed K. Thus $||\mu||_\infty < 1$. Therefore $[\mu] = \lim[\mu_n]$ in $(T(M), d_T)$, which proves (c). Since f_n and f^{μ_n} are normalized to fix $0, 1$, and ∞, the families (f_n) and (f^{μ_n}) are normal families. Thus (c) and (d) follow by passing, if necessary, to subsequences and using Lemma 5.2.1, which says that if $\mu_n \to \mu$ almost everywhere, then $f^{\mu_n} \to f^\mu$ uniformly on compact subsets. □

This immediately gives the following theorem.

Theorem 7.3.4. *The Teichmüller space $(T(M), d_T)$ is complete.*

Theorem 7.3.5. *The Teichmüller spaces of quasiconformally equivalent Riemann surfaces are isometrically bijective.*

Proof. Let $h : M \to N$ be a quasiconformal mapping between Riemann surfaces. The mapping $f \to f \circ h^{-1}$ is a bijection of the family of all quasiconformal mappings of M onto the family of all quasiconformal mappings of N. If $w_i = f_i \circ h^{-1}$ for $i = 1, 2$, then

$$w_2 \circ w_1^{-1} = f_2 \circ f_1^{-1}.$$

Thus f_1 and f_2 determine the same point of $T(M)$ if and only if w_1 and w_2 determine the same point of $T(N)$. Therefore

$$[f]_{T(M)} \to [f \circ h^{-1}]_{T(N)}$$

is a bijective mapping of $T(M)$ onto $T(N)$ and is clearly an isometry. □

Using this theorem, we can map $[h] \in T(M)$ to $[0] \in T(M)$ and so anything that can be proved at the origin of Teichmüller space can be transported to arbitrary points of $T(M)$.

Remark. If M and N are compact Riemann surfaces of the same genus, then they are homeomorphic and by Theorem 7.2.1 are quasiconformally equivalent. Now Theorem 7.3.5 shows that the Teichmüller spaces of compact Riemann surfaces of the same genus are isometrically bijective.

Exercise 7.3.1. As noted at the beginning of the chapter, $T(M)$ is a group under the obvious composition $[f] \circ [g] = [f \circ g]$. However, show that $T(M)$ is not a topological group, that is, show that the group composition operator is not continuous with respect to the Teichmüller metric. This fact is proved in [23]. (Hints: Let M be the upper half-plane $\mathbb{H}$ and model Teichmüller space on the space of quasisymmetric maps on $\mathbb{R}$, normalized to fix three points, say $-1, 1,$

and ∞. Let f_n be normalized quasisymmetric maps such that $d_T(f_n, I) \to 0$, where I is the identity, and such that a quasiconformal extension of f_n to $\mathbb{H}$ is asymptotically conformal (recalling Definition 7.2.2). Now let $g(z) = z|z|$ on $\mathbb{H}$, which is clearly not asymptotically conformal. Show that $d_T(g \circ f_n, g)$ does not converge to zero. Note, however, that the other way of composing is continuous, so that $d_T(f_n \circ g, g) \to 0$.)

7.4 The Teichmüller space of a torus

As an extended example, we show that the Teichmüller space of a torus can be identified with the upper half-plane $\mathbb{H}$. The remark after Theorem 7.3.5 shows that Teichmüller spaces of any two tori are isometrically bijective. Now, the covering group G of $\mathbb{C}$ over a torus M consists of translations

$$z \mapsto z + m\omega_1 + n\omega_2$$

where $\omega_1, \omega_2 \in \mathbb{C}$ and linearly independent over $\mathbb{R}$, and $m, n \in \mathbb{Z}$. The pair (ω_1, ω_2) is called a base of G, and is called normalized if

$$\operatorname{Im}\left(\frac{\omega_1}{\omega_2}\right) > 0.$$

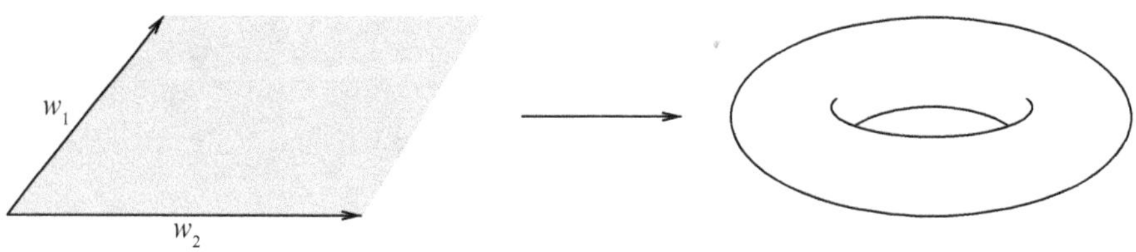

Lemma 7.4.1. *Let (ω_1, ω_2) be a base of G. Then (ω_1', ω_2') is a base of G if and only if*

$$\omega_1' = a\omega_1 + b\omega_2, \qquad \omega_2' = c\omega_1 + d\omega_2 \tag{7.4}$$

where $a, b, c, d \in \mathbb{Z}$ and $|ad - bc| = 1$. If (ω_1, ω_2) is normalized, then (ω_1', ω_2') is normalized if and only if $ad - bc = 1$.

Proof. The condition is clearly necessary, and is sufficient if (7.4) can be solved for ω_1, ω_2 such that ω_1, ω_2 are linear combinations of ω_1', ω_2' with coefficients in $\mathbb{Z}$. This occurs if $|ad - bc| = 1$. Now,

$$\operatorname{Im}\left(\frac{\omega_1'}{\omega_2'}\right) = \frac{ad - bc}{|c\omega_1/\omega_2 + d|^2} \operatorname{Im}\left(\frac{\omega_1}{\omega_2}\right)$$

and so if (ω_1, ω_2) is normalized, then we have that (ω_1', ω_2') is normalized if and only if $ad - bc = 1$. □

The elliptic modular group Γ is the restriction to $\mathbb{H}$ of the group of all Möbius transformations of the form

$$z \mapsto \frac{az+b}{cz+d},$$

where $a, b, c, d \in \mathbb{Z}$ and $ad - bc = 1$. The group Γ contains elliptic transformations, for example

$$z \mapsto -\frac{1}{z}$$

is of order 2 and

$$z \mapsto \frac{-1}{(z+1)}$$

is of order 3, but can be shown to act properly discontinuously on $\mathbb{H}$. Thus $\mathbb{H}/\Gamma$ can be considered as a Riemann surface.

Now, let $M = \mathbb{C}/G$ be a torus with (ω_1, ω_2) a normalized base of G. From Lemma 7.4.1, if (ω_1', ω_2') is a pair of non-zero complex numbers, then $(\alpha\omega_1', \alpha\omega_2')$ is a normalized base for G for $\alpha \neq 0$ if and only if ω_1'/ω_2' is equivalent to ω_1/ω_2 under the action of Γ. In this way, every torus can be represented as a point of $\mathbb{H}$.

Lemma 7.4.2. *Let M be a torus and $p, q \in M$. There exists a conformal map $f : M \to M$ homotopic to the identity such that $f(p) = q$.*

Proof. Let $\pi : \mathbb{C} \to M \simeq \mathbb{C}/G$ be the projection and $z \in \pi^{-1}(p)$, $w \in \pi^{-1}(q)$. Note that a translation commutes with every $g \in G$. Therefore

$$\xi \mapsto \xi + t(w - z)$$

can be projected to a conformal mapping $f_t : M \to M$ for every t with $0 \leq t \leq 1$. This mapping gives the required homotopy. □

This lemma means that we can restrict ourselves to looking at normalized lifts $f : \mathbb{C} \to \mathbb{C}$ of mappings between tori $\varphi : M \to N$ such that $f(0) = 0$. Then the elements of the covering group of $\mathbb{C}$ over M,

$$z \mapsto z + m\omega_1 + n\omega_2$$

are mapped by f to the elements

$$z \mapsto mf(\omega_1) + nf(\omega_2)$$

of the covering group of $\mathbb{C}$ over N.

Theorem 7.4.1. *Let $M = \mathbb{C}/G$ and $N = \mathbb{C}/G'$ be tori and $\theta : G \to G'$ an isomorphism. Then there exists a homeomorphism of M onto N inducing θ.*

Proof. Let (ω_1, ω_2) be a base of G and

$$(\omega_1, \omega_2) \mapsto (\omega_1', \omega_2')$$

under θ. Consider the affine map λ fixing 0 and mapping ω_i to ω_i' for $i = 1, 2$. Then λ determines θ and projects to a homeomorphism of M onto N. □

Now let $\varphi : M \to N$ be a homeomorphism with normalized lift f such that $f(\omega_i) = \omega_i'$ for $i = 1, 2$. If $h = f \circ \lambda^{-1}$, then

$$h(z + \omega_i') = h(z) + \omega_i'$$

for $i = 1, 2$. Therefore h is sense-preserving and consequently f and λ are simultaneously sense-preserving. Let $\tau = \omega_1/\omega_2$ and $\tau' = \omega_1'/\omega_2'$. If $\tau' \neq \overline{\tau}$, then we have $\lambda(z) = \alpha(z + \mu\overline{z})$, and a calculation shows that

$$|\mu| = \left| \frac{\tau' - \tau}{\tau' - \overline{\tau}} \right| .$$

Suppose (ω_1, ω_2) is a normalized base of G with $\text{Im}(\tau) > 0$. Then $\text{Im}(\tau') > 0$ if and only if $|\mu| < 1$. However $|\mu| < 1$ if and only if λ is sense-preserving. Therefore φ is sense-preserving if and only if

$$\text{Im}\left(\frac{f(\omega_1)}{f(\omega_2)}\right) > 0.$$

In the case where $\tau' = \overline{\tau}$, then φ is sense-reversing and

$$\text{Im}\left(\frac{f(\omega_1)}{f(\omega_2)}\right) < 0.$$

If $|\mu| < 1$, then $\lambda(z) = \alpha(z + \mu\overline{z})$ is quasiconformal. Since the projection of a quasiconformal map of the plane is a quasiconformal map of the torus, all tori are quasiconformally equivalent.

Theorem 7.4.2. *Let $M \simeq \mathbb{C}/G$ and $N \simeq \mathbb{C}/G'$ be tori and let (ω_1, ω_2) and (ω_1', ω_2') respectively be normalized bases of G and G'. Then M and N are conformally equivalent if and only if ω_1/ω_2 and ω_1'/ω_2' are equivalent under the elliptic modular group Γ.*

Proof. If M and N are conformally equivalent, there is a conformal map of M onto N whose lift $f : \mathbb{C} \to \mathbb{C}$ is normalized. Since f is a conformal self-mapping of $\mathbb{C}$ with $f(0) = 0$ and $f(z) = \alpha z$, then $f(\omega_i) = \alpha\omega_i$ for $i = 1, 2$ and so $(\alpha\omega_1, \alpha\omega_2)$ is a base of G'.

Conversely, if $(\alpha\omega_1, \alpha\omega_2)$ is a base of G', the projection of $z \mapsto \alpha z$ is a conformal mapping of M onto N. Now $(\alpha\omega_1, \alpha\omega_2)$ is a base of G' if and only if ω_1/ω_2 and ω_1'/ω_2' are equivalent under Γ. □

Let $N \simeq \mathbb{C}/G'$ and $\theta : G \to G'$ be a given isomorphism. If θ is induced by a sense-preserving homeomorphism $\varphi : M \to N$ with normalized lift f, then $f(\omega_1)$ and $f(\omega_2)$ do not depend on the choice of φ. The hyperbolic distance in $\mathbb{H}$ between ω_1/ω_2 and $f(\omega_1)/f(\omega_2)$ depends only on θ, not the choice of base for G. This distance is denoted by δ_θ and measures how much θ deviates from a conformal mapping. We state the following lemma without proof, which can be found in [31].

Lemma 7.4.3. *Let $\theta : G \to G'$ be an isomorphism generated by a normalized K-quasiconformal mapping f. Then*

$$\delta_\theta \leq \frac{1}{2} \log K$$

with equality if and only if f is the affine translation generating θ.

Let $p \in T(M)$ be a point in the Teichmüller space of the torus M. Considering representatives of p, we can just use mappings φ of M with a normalized lift f. Set

$$\psi(p) = \frac{f(\omega_1)}{f(\omega_2)}.$$

We show that this is a well-defined mapping of $T(M)$. Let $\widetilde{\varphi} \in p$ with a normalized lift $\widetilde{f}$. Then there exists a conformal map $\sigma : \widetilde{\varphi}(M) \to \varphi(M)$ with a normalized lift such that $\sigma \circ \widetilde{\varphi}$ is homotopic to φ. The lifts of φ and $\sigma \circ \widetilde{\varphi}$ induce the same group isomorphism and the lift of σ is of the form $z \mapsto \alpha z$, and so

$$f(\omega_i) = \alpha \widetilde{f}(\omega_i)$$

for $i = 1, 2$ and so $f(\omega_i)/f(\omega_2)$ is independent of the choice of $\varphi \in p$. Note that $\psi(p) \in \mathbb{H}$.

Theorem 7.4.3. *The mapping $\psi : T(M) \to \mathbb{H}$ defined by*

$$\psi([\varphi]) = \frac{f(\omega_1)}{f(\omega_2)}$$

where f is the normalized lift of φ, is a bijective isometry of $T(M)$ onto $\mathbb{H}$ with the hyperbolic metric.

Proof. Firstly we show that ψ is injective. If $\psi(p_1) = \psi(p_2)$, then there exist conformal mappings $\varphi_i \in p_i$ for $i = 1, 2$ whose normalized lifts f_i satisfy

$$f_1(\omega_i) = \alpha f_2(\omega_i).$$

If $s(z) = \alpha z$, then f_1 and $s \circ f_2$ determine the same group isomorphism and therefore the projections φ_1 and $\sigma \circ \varphi_2$ are homotopic. Since σ is conformal, then $p_1 = p_2$.

To show that ψ is surjective, given $z \in \mathbb{H}$, choose $\omega_2' \in \mathbb{C}$ and set $\omega_1' = z\omega_2'$. By Theorem 7.4.1, there exists a homeomorphism φ of M with normalized lift f such that $f(\omega_i) = \omega_i'$ for $i = 1, 2$. Now φ is sense-preserving and determines $p \in T(M)$, and also $\psi(p) = z$.

Finally, we show that ψ is an isometry. Given $p_1, p_2 \in T(M)$, consider representatives φ_i with normalized lifts f_i. Let $\varphi : \varphi_1(M) \to \varphi_2(M)$ be the extremal in the class of quasiconformal mappings homotopic to $\varphi_2 \circ \varphi_1^{-1}$, and having a normalized lift. Now

$$d_T(p_1, p_2) = \frac{1}{2} \log K$$

where K is the maximal dilatation of φ. However, if f is the normalized lift of φ, then $1/2 \log K$ is the hyperbolic distance between the points $f_1(\omega_1)/f_1(\omega_2)$ and $f(f_1(\omega_1))/f(f_1(\omega_2))$. Now $f \circ f_1$ induces the same group isomorphism as f_2 and so $f(f_1(\omega_i)) = f_2(\omega_i)$ for $i = 1, 2$. Thus

$$d_T(p_1, p_2) = d_h(\psi(p_1), \psi(p_2)).$$

□

This theorem shows that the Teichmüller space of a torus can be identified with the upper half-plane $\mathbb{H}$.

7.5 Schwarzian derivatives and quadratic differentials

If f is holomorphic in a domain Ω and $f'(z) \neq 0$ in Ω, then the Schwarzian derivative of f is

$$S_f = \left(\frac{f''}{f'}\right)' - \frac{1}{2}\left(\frac{f''}{f'}\right)^2. \tag{7.5}$$

If also $f(z) \neq 0$ in Ω, then direct computation shows

$$S_f(z) = S_{1/f}(z)$$

and this formula shows how to define the Schwarzian derivative of a meromorphic function at simple poles. Thus the Schwarzian derivative can be defined for locally injective meromorphic functions, and S_f is itself holomorphic. Let

$$A(z) = \frac{az + b}{cz + d}$$

be a Möbius transformation, then differentiating gives

$$\frac{A''(z)}{A'(z)} = -\frac{2c}{cz + d}$$

and

$$\left(\frac{A''(z)}{A'(z)}\right)' = \frac{2c^2}{(cz + d)^2}$$

from which we see that

$$S_A = 0.$$

Conversely, starting with the equation $S_f = 0$ and setting $g = f''/f'$, then (7.5) gives $g' = g^2/2$. Solving this differential equation shows that every solution of $S_f = 0$ is a Möbius transformation. Schwarzian derivatives satisfy the composition rule

$$S_{f \circ g} = (S_f \circ g)g'^2 + S_g \tag{7.6}$$

and so if g is a Möbius transformation and $S_g = 0$ then

$$S_{f\circ g} = (S_f \circ g)g'^2. \tag{7.7}$$

On the other hand, if f is a Möbius transformation, then

$$S_{f\circ g} = S_g. \tag{7.8}$$

To define the Schwarzian derivative at ∞, assume that f is locally injective and meromorphic in a neighbourhood of ∞, then $h(z) = f(1/z)$ is defined in a neighbourhood of 0. Using (7.7),

$$z^4 S_h(z) = S_f\left(\frac{1}{z}\right)$$

and so we can define

$$S_f(\infty) = \lim_{z\to 0} z^4 S_h(z)$$

and S_f is holomorphic at ∞. Thus the Schwarzian derivative can be defined for a locally injective meromorphic function f on any domain Ω. The following theorem shows that the Schwarzian derivative can be prescribed.

Theorem 7.5.1. *Let g be a holomorphic function in a simply connected domain Ω. Then there is a meromorphic function f in Ω such that*

$$S_g = f \tag{7.9}$$

which is unique up to an arbitrary Möbius transformation.

Proof. The existence of such an f follows from standard theory on differential equations, which we omit. For the uniqueness, from (7.8), if f is a solution of (7.9) and A is a Möbius transformation, then $A \circ f$ is also a solution. Conversely, if f_1 and f_2 are solutions of (7.9) in Ω, then $f_2 \circ f_1^{-1}$ can be defined locally, since f_1 is locally injective. Since $S_f = S_g$, then $S_{g\circ f^{-1}} = 0$ from (7.6), which means $g = A \circ f$ for a Möbius transformation A. But we then have that $g = A \circ f$ everywhere in Ω, which proves the uniqueness. □

We now move on to consider spaces of quadratic differentials. Consider first the space $Q(\mathbb{H})$ of all functions holomorphic in $\mathbb{H}$ which have finite norm given by

$$||\varphi||_{Q(\mathbb{H})} = \sup_{z\in\mathbb{H}} \rho_{\mathbb{H}}^{-2}(z)|\varphi(z)| < \infty,$$

where $\rho_{\mathbb{H}}$ is the hyperbolic density on $\mathbb{H}$ given by

$$\rho_{\mathbb{H}}(z) = (2y)^{-1}$$

and $z = x + iy$. This norm is called the Bers norm and $Q(\mathbb{H})$ is the Bers space of $\mathbb{H}$. The space $Q(\mathbb{H})$ is a linear space over $\mathbb{C}$.

Proposition 7.5.1. *$Q(\mathbb{H})$ is complete and thus a Banach space.*

Proof. If (φ_n) is a Cauchy sequence in $Q(\mathbb{H})$, then $\rho_{\mathbb{H}}^{-2}\varphi_n$ is a Cauchy sequence in L^∞. Since L^∞ is complete, there exists $\psi \in L^\infty$ such that $\rho_{\mathbb{H}}^{-2}\varphi_n \to g$ in L^∞. Since $\rho_{\mathbb{H}}$ is locally bounded, φ_n converges locally uniformly to $\varphi = \rho_{\mathbb{H}}^2\psi$. This implies that φ is holomorphic, $||\varphi - \varphi_n||_{Q(\mathbb{H})} \to 0$ and $\varphi \in Q(\mathbb{H})$. □

Note that by Theorem 7.5.1, every $\varphi \in Q(\mathbb{H})$ is the Schwarzian derivative of some meromorphic function f in $\mathbb{H}$.

Now let G be a group acting on $\mathbb{H}$. Then $Q(G)$ is the space of all functions φ holomorphic in $\mathbb{H}$ which are quadratic differentials for G, that is, for all $g \in G$,

$$(\varphi \circ g)g'^2 = \varphi$$

which have finite norm given by

$$||\varphi||_{Q(G)} = \sup_{z\in\mathbb{H}} \rho_{\mathbb{H}}^{-2}(z)|\varphi(z)| < \infty. \tag{7.10}$$

The expression in (7.10), for $\varphi \in Q(G)$, is invariant under G, which means that the supremum in (7.10) can be taken over an arbitrary Dirichlet region $D \subset \mathbb{H}$ of G,

$$||\varphi||_{Q(G)} = \sup_{z\in D} \rho_{\mathbb{H}}^{-2}(z)|\varphi(z)| < \infty.$$

We can similarly define $Q(M)$ for a Riemann surface (or general plane domain) M by using the norm

$$||\widetilde{\varphi}||_{Q(M)} = \sup_{p\in M} \rho_M^{-2}|\widetilde{\varphi}(p)|,$$

where ρ_M is the hyperbolic density on M, and noting that $\rho_M^{-2}|\widetilde{\varphi}(p)|$ is a function on M, whereas $|\varphi|$ is usually not. If $M \simeq \mathbb{H}/G$ with $\pi_M : D \to M$ the covering map, then there is a one-to-one correspondence between $Q(M)$ and $Q(G)$ given by

$$\widetilde{\varphi} \circ \pi_M \longleftrightarrow \varphi$$

by using the invariance properties of ρ_M, $\rho_{\mathbb{H}}$, $\widetilde{\varphi}$ and φ under G. In this way we can consider $Q(G)$ instead of $Q(M)$. Clearly $Q(G) \subseteq Q(\mathbb{H})$, and let $\varphi_n(z) \in Q(G)$ converge to $\varphi \in Q(\mathbb{H})$. Given $g \in G$, $\varphi_n(g(z)) \to \varphi(g(z))$ locally uniformly on $\mathbb{H}$. Then

$$\varphi(g(z))g'(z)^2 = \lim_{n\to\infty} \varphi_n(g(z))g'(z)^2 = \lim_{n\to\infty} \varphi_n(z) = \varphi(z)$$

and so $\varphi \in Q(G)$. This shows that $Q(G)$ is a closed subspace of $Q(\mathbb{H})$ and therefore a Banach space. The Bers norm induces a metric on $Q(G)$, given by

$$d_{Q(G)}(\varphi_1, \varphi_2) = ||\varphi_1 - \varphi_2||_{Q(G)}$$

for $\varphi_1, \varphi_2 \in Q(G)$.

Now let f_1 and f_2 be meromorphic functions on a domain Ω and let $h : \Omega' \to \Omega$ be conformal. Using the invariance property (7.6),

$$S_{f_1\circ h} - S_{f_2\circ h} = (S_{f_1} \circ h - S_g \circ h)h'^2.$$

Since the hyperbolic density is conformally invariant, $\rho_{\Omega'} = (\rho_\Omega \circ h)|h'|$, then by writing $w = h(z)$, we have the invariance formula

$$\frac{|S_{f_1}(w) - S_{f_2}(w)|}{\rho^2_\Omega(w)} = \frac{|S_{f_1 \circ h}(z) - S_{f_2 \circ h}(z)|}{\rho^2_{\Omega'}(z)}. \tag{7.11}$$

Combining this with (7.10) gives

$$||S_{f_1} - S_{f_2}||_{Q(\Omega)} = ||S_{f_1 \circ h} - S_{f_2 \circ h}||_{Q(\Omega')}.$$

If $f_2 = h^{-1}$ is a conformal mapping of Ω, then

$$||S_{f_1} - S_{f_2}||_{Q(\Omega)} = ||S_{f_1 \circ f_2^{-1}}||_{Q(f_2(\Omega))}.$$

In the special case of f_1 being the identity,

$$||S_{f_2}||_{Q(\Omega)} = ||S_{f_2^{-1}}||_{Q(f_2(\Omega))}.$$

Lastly, if $f_2 = h^{-1}$ is a Möbius transformation, then

$$||S_{f_1}||_{Q(\Omega)} = ||S_{f_1 \circ f_2^{-1}}||_{Q(f_2(\Omega))}, \tag{7.12}$$

which shows that $||S_f||_{Q(\Omega)}$ is completely invariant with respect to Möbius transformations.

Example 7.5.1. If f is a conformal mapping of $\mathbb{D}$, we will show that

$$||S_f||_{Q(\mathbb{D})} \le 6, \tag{7.13}$$

and the bound can be attained. Note that by (7.12), f can be defined on any disk and still obey (7.13). Fix $z_0 \in \mathbb{D}$ and we will estimate

$$|S_f(z_0)|\rho_\mathbb{D}(z_0)^{-2} = \frac{(1 - |z_0|^2)^2}{4}|S_f(z_0)|.$$

This expression is invariant under Möbius transformations, so we can assume that $z_0 = 0$. Since f can be replaced by $A \circ f$, where A is any Möbius transformation, f can be assumed to be normalized by $f'(0) = 1$. Let a_n denote the nth power series coefficient of f. The function

$$\frac{1}{f(1/z)} = z + \sum_{n=0}^{\infty} b_n z^{-n}$$

satisfies the conditions of the classical Area theorem, see for example [42], and so we can conclude that

$$|b_1| = |a_2^2 - a_3| \le 1.$$

However, $S_f(0) = 6(a_3 - a_2^2)$, and so $|S_f(0)| \le 6$ from which (7.13) follows. For any Koebe function

$$f(z) = \frac{z}{(1 + e^{i\theta} z)^2}$$

the coefficient of b_1 has absolute value 1, and so equality holds for (7.13). More generally, equality holds for $A \circ f \circ B$, where B is a conformal self-mapping of $\mathbb{D}$,

f is a Koebe function and A is an arbitrary Möbius transformation. For example, $f(z) = z^2$ in $\mathbb{H}$ satisfies $||S_f||_{Q(\mathbb{H})} = 3/2$.

Recalling the quasiconformal mappings f^μ and f_μ, we now switch the definition of f_μ so that f_μ has complex dilatation equal to 0 on $\mathbb{H}$ and $\mu(\overline{z})$ on $\mathbb{H}^-$. This has the effect that $f_\mu|_{\mathbb{H}}$ is conformal, whereas before $f_\mu|_{\mathbb{H}^-}$ was conformal. Now, since $f_\mu|_{\mathbb{H}^-}$ is a conformal map, the set of such maps characterizes $T(M)$. If $M \simeq \mathbb{H}/G$, then for every $g \in G$, $f_\mu \circ g \circ f_\mu^{-1}$ is a Möbius transformation. Therefore, using the transformation rules (7.7) and (7.8), we have

$$S_{f_\mu|_{\mathbb{H}}} = S_{(f_\mu \circ g \circ f_\mu^{-1}) \circ f_\mu|_{\mathbb{H}}} = S_{f_\mu \circ g|_{\mathbb{H}}} = (S_{f_\mu|_{\mathbb{H}}} \circ g) g'^2. \tag{7.14}$$

This shows that the Schwarzian derivative is a quadratic differential for the group G acting on $\mathbb{H}$, and its projection is a holomorphic quadratic differential on $\mathbb{H}/G$, which is the mirror image of the Riemann surface M, since we have changed the definition of f_μ.

Lemma 7.5.1. *The following conditions are equivalent.*

(a) $S_{f_\mu|_{\mathbb{H}}}$ *is a quadratic differential for* G,

(b) $f_\mu \circ g \circ f_\mu^{-1}$ *agrees with a Möbius transformation in* $f_\mu(\mathbb{H})$ *for* $g \in G$,

(c) $f^\mu \circ g \circ (f^\mu)^{-1}$ *agrees with a Möbius transformation on* $\mathbb{R}$ *for* $g \in G$.

Proof. The equivalence of (a) and (b) follows from (7.14). Now assmue (b) holds, so $A = f_\mu \circ g \circ f_\mu^{-1}$ in $f_\mu(\mathbb{H})$ is a Möbius transformation, then

$$B = f^\mu \circ f_\mu^{-1} \circ A \circ f_\mu \circ (f^\mu)^{-1} : \mathbb{H}^- \to \mathbb{H}^-$$

is a Möbius transformation which agrees with $f^\mu \circ g \circ (f^\mu)^{-1}$ on $\mathbb{R}$, and so (c) holds.

Conversely, assume that (c) holds, so that $A = f^\mu \circ g \circ (f^\mu)^{-1}$ is a Möbius transformation on $\mathbb{R}$. Let

$$B = f_\mu \circ (f^\mu)^{-1} \circ A \circ f^\mu \circ f_\mu^{-1}$$

on the closure of $f_\mu(\mathbb{H}^-)$, and $B = f_\mu \circ g \circ f_\mu^{-1}$ in $f_\mu(\mathbb{H})$. Then B is a homeomorphism of $\mathbb{C}$ which is conformal in $f_\mu(\mathbb{H}^-)$ and $f_\mu(\mathbb{H})$. Hence B is a Möbius transformation, and (b) holds, which completes the proof. □

7.6 The Bers embedding

We have the mapping

$$\mu \mapsto S_{f_\mu|_{\mathbb{H}}}$$

which maps the open unit ball $B(M)$ of $L^\infty(M)$ into the space of quadratic differentials $Q(M)$. This induces a mapping

$$\lambda_M : T(M) \to Q(M),$$

which is called the Bers embedding, and the aim of this section is to explore the properties of this mapping.

The first thing to note is that if two holomorphic functions f_1 and f_2 on $\mathbb{H}$ have the same Schwarzian derivative, then by Theorem 7.5.1, one is equal to the other postcomposed by a Möbius transformation. Now since both functions are normalized at three points, then f_1 and f_2 must be identical on $\mathbb{H}$, and therefore determine the same Teichmüller class. This shows that the mapping λ_M is one-to-one onto its image.

By transferring equation (7.13) to $\mathbb{H}$ via a conformal mapping, which does not change the norm of a Schwarzian derivative, we see that for any holomorphic mapping f of $\mathbb{H}$ satisfies

$$||S_f||_{Q(\mathbb{H})} \leq 6,$$

and so the ball of radius 6 centred at the origin of $Q(M)$ contains the image of λ_M. The following lemma shows that the ball of radius 2 centred at the origin of $Q(M)$ is contained in the image of λ_M.

Lemma 7.6.1. *Let φ be a holomorphic function in $\mathbb{H}$ satisfying*

$$||\varphi(z)y^2||_\infty < \frac{1}{2},$$

where $z = x + iy$. Let $\mu(z) = -2y^2\varphi(\overline{z})$ for $z \in \mathbb{H}^-$ and $\mu(z) = 0$ for $z \in \mathbb{H}$. Then

$$S_{f_\mu|_{\mathbb{H}}}(z) = \varphi(z).$$

Additionally, if φ is a quadratic differential for the Fuchsian group G, then μ is a Beltrami differential for G.

Proof. Initially, we assume that φ is holomorphic in a neighbourhood of $\mathbb{H}$, and that $|\varphi(z)| = O(|z|^{-4})$ as $z \to \infty$. Let η_1 and η_2 be two linearly independent solutions of

$$\eta'' = -\frac{1}{2}\varphi\eta$$

normalized by $\eta_1'\eta_2 - \eta_2'\eta_1 = 1$, and form

$$\widehat{f}(z) = \begin{cases} \dfrac{\eta_1(\overline{z}) + (z-\overline{z})\eta_1'(\overline{z})}{\eta_2(\overline{z}) + (z-\overline{z})\eta_2'(\overline{z})}, & \text{for } z \in \mathbb{H}^-, \\[2ex] \dfrac{\eta_1(z)}{\eta_2(z)}, & \text{for } z \in \mathbb{H}. \end{cases}$$

Since η_1 and η_2 are holomorphic along $\mathbb{R}$, the two formulas for $\widehat{f}$ in $\mathbb{H}$ and $\mathbb{H}^-$ are continuous and coincide for $z \in \mathbb{R}$. In both formulas, the numerators and denominators do not vanish simultaneously due to the normalization

$$\eta_1'\eta_2 - \eta_2'\eta_1 = 1.$$

By direct calculation,

$$\frac{\widehat{f}_{\overline{z}}}{\widehat{f}_z} = -2y^2\varphi(\overline{z})$$

for $z \in \mathbb{H}^-$, and $\widehat{f}_{\overline{z}} = 0$ for $z \in \mathbb{H}$. Since $\widehat{f}$ has Beltrami coefficient μ with $||\mu||_\infty < 1$, it follows that $\widehat{f}$ is quasiconformal.

In order to show that $\widehat{f}$ is a homeomorphism of $\widehat{\mathbb{C}}$ onto $\widehat{\mathbb{C}}$, it suffices, due to topological reasons, to show that it is a homeomorphism at each point. Since $J_{\widehat{f}} = |\widehat{f}_z|^2(1 - |\mu|^2)$, and $||\mu||_\infty < 1$, it is enough to show that $\widehat{f}_z \neq 0$. Now, in $\mathbb{H}$, we have

$$\widehat{f}_z = \frac{1}{\eta_2(z)^2}$$

and since η_2 is holomorphic in $\mathbb{H}$, $\widehat{f}_z$ cannot be zero. If $\eta_2(z) = 0$, then consider the map $1/\widehat{f}$. Then, in $\mathbb{H}$, we have

$$\left(\frac{1}{\widehat{f}}\right)_z = (-\widehat{f}^{-1})\widehat{f}_z = \frac{1}{\eta_1(z)^2}.$$

Since η_1 and η_2 cannot be simultaneously zero, either $\widehat{f}$ or $1/\widehat{f}$ is a local homeomorphism at every point. Thus $\widehat{f}$ is a local homeomorphism at every point of $\mathbb{H}$.

If $z \in \mathbb{H}^-$, we get

$$\widehat{f}_z = (\eta_2(\overline{z}) + (z - \overline{z})\eta_2'(\overline{z}))^{-2}.$$

Thus $\widehat{f}_z$ cannot be zero, and considering $1/\widehat{f}$ deals with the case where $\widehat{f}_z$ is ∞. Since η_1 and η_2 are holomorphic on $\mathbb{R}$, and the two formulas for $\widehat{f}$ match along $\mathbb{R}$, then $\widehat{f}_z \neq 0$ on $\mathbb{C}$.

It remains to show that $\widehat{f}$ is a local homeomorphism at ∞. Let $t = 1/z$ be a local parameter at ∞. We have to show that

$$\widehat{f}_t = \frac{-\widehat{f}_z}{t^2}$$

approaches a non-zero limit as $t \to 0$. Since $|\varphi| = O(|z|^{-4})$ as $z \to \infty$, the solution $\eta_2(z)$ will be of the form $a_2 z + b_2 + O(|z|^{-1})$. Therefore $\eta_2'(z) = a_2 + O(|z|^{-2})$ and

$$\widehat{f}_z = (a_2\overline{z} + b_2 + a_2 - a_2\overline{z} + O(|z|^{-1}))^{-2}$$

and

$$\widehat{f}_t = -(a_2 + b_2 t + O(t))^{-2}.$$

This shows that $\widehat{f}$ is a local homeomorphism at ∞ as long as $a_2 \neq 0$. If $a_2 = 0$, then we consider $1/\widehat{f}$ and we get

$$\lim_{t\to 0}(\widehat{f}^{-1})_t = -a_1^{-2},$$

where $\eta_1(z) = a_1 z + b_1 + O(|z|^{-1})$. Now the normalization $\eta_1'\eta_2 - \eta_2'\eta_1 = 1$ implies that $a_1 b_2 - a_2 b_1 = 1$, and so not both a_1 and a_2 are zero. Hence $\widehat{f}$ is a local homeomorphism in both cases.

Now, in order to remove the hypothesis that φ is analytic on $\mathbb{R}$ with a fourth-order zero at ∞, take Möbius transformations A_n such that the closure of $A_n(\mathbb{H})$ is compact in $\mathbb{H}$ and such that $A_n(z) \to z$ as $n \to \infty$. Consider

$$\varphi_n(z) = \varphi(A_n(z))(A_n'(z))^2.$$

We see that $|\varphi_n(z)| = O(|z|^{-4})$ as $n \to \infty$, and that φ_n is holomorphic in a neighbourhood of $\mathbb{R}$. Since $A_n(\mathbb{H}) \subset \mathbb{H}$, we have

$$\rho_{\mathbb{H}}(A_n(z))|A_n'(z)| \le \rho_{\mathbb{H}}(z),$$

where $\rho_{\mathbb{H}}|dz|$ is the hyperbolic metric for $\mathbb{H}$. Hence $|A_n'(z)||y| < |\mathrm{Im}(A_n(z))|$ and then

$$|y^2\varphi_n(z)| = |\varphi(A_n(z))A_n'(z)^2|y^2 < |\varphi(A_n(z))||\mathrm{Im}(A_n(z))|^2$$

which implies

$$||y^2\varphi_n(z)||_\infty \le ||y^2\varphi(z)||_\infty.$$

Now $\varphi_n \to \varphi$ uniformly on compact subsets of $\mathbb{H}$, and thus normalized solutions to

$$\eta'' = -\frac{1}{2}\varphi_n\eta$$

will converge uniformly on compact subsets of $\mathbb{H}$ to normalized solutions of

$$\eta'' = -\frac{1}{2}\varphi\eta.$$

Form $\widehat{f_n}$ corresponding to φ_n. Each $\widehat{f_n}$ is a quasiconformal homeomorphism of $\widehat{\mathbb{C}}$ normalized at 0, 1 and ∞, and with dilatation bounded by

$$||y^2\varphi(z)||_\infty < 1.$$

By Theorem 5.2.3, the existence theorem for the Beltrami equation, we can form $\widehat{f}$, a normalized quasiconformal homeomorphism of $\widehat{\mathbb{C}}$ with Beltrami coefficient $-1/2y^2\varphi(\overline{z})$ in $\mathbb{H}^-$ and 0 in $\mathbb{H}$. Since $y^2\varphi_n(\overline{z})$ converges pointwise to $y^2\varphi(z)$, $\widehat{f_n}$ converges locally uniformly to $\widehat{f}$. Furthermore, $S_{\widehat{f_n}|_{\mathbb{H}}} = \varphi_n$ converges locally uniformly to $S_{f|_{\mathbb{H}}}$ in $\mathbb{H}$, and hence $S_{f|_{\mathbb{H}}} = \varphi$. It is easy to show that if φ is quadratic differential for G, then $y^2\varphi(\overline{z})$ is a Beltrami differential for G, and is left to the reader. □

Fix S_f in the image of the universal Teichmüller space $T(\mathbb{H})$ under $\lambda = \lambda_{\mathbb{H}}$. For $S_h \in Q(\mathbb{H})$, consider the function $g = h \circ f^{-1}$, which is meromorphic in the quasidisk $f(\mathbb{H})$. Using a theorem in Lehto, there exists $\epsilon > 0$ small enough so that if $||S_g||_{Q(f(\mathbb{H}))} < \epsilon$, then g is univalent in $f(\mathbb{H})$ and has quasiconformal extension to $\mathbb{C}$. So now choosing $S_h \in Q(\mathbb{H})$ such that $||S_h - S_f||_{Q(\mathbb{H})} < \epsilon$, then

$$||S_g||_{Q(f(\mathbb{H}))} = ||S_h - S_f||_{Q(\mathbb{H})} < \epsilon$$

and since $h = g \circ f$, this means that S_h is in the image of $T(\mathbb{H})$ under λ. The following proposition generalizes this to arbitrary Bers embeddings.

Proposition 7.6.1. *The image of $T(M)$ under λ_M in $Q(M)$ is open.*

Proof. Writing $s_\mu = S_{f_\mu|_{\mathbb{H}}}$, let $\psi \in T(\mathbb{H})$ be given by

$$\psi(s_\nu) = s_{\widetilde{\alpha_\mu}(\nu)}$$

where

$$f^{\widetilde{\alpha_\mu}(\nu)} = f^\nu \circ (f^\mu)^{-1}.$$

If α_μ is the mapping induced from $\widetilde{\alpha_\mu}$, given by

$$\alpha_\mu([\nu]) = [\widetilde{\alpha_\mu}(\nu)]$$

and λ is the Bers embedding of universal Teichmüller space, then

$$\psi = \lambda \circ \alpha_\mu \circ \lambda^{-1}.$$

Therefore ψ is a homeomorphism of $T(\mathbb{H})$ onto itself, and ψ maps $\lambda_M(T(M))$ onto $\lambda_{M^\mu}(T(M^\mu))$, where M^μ is the Riemann surface quasiconformally equivalent to $M \simeq \mathbb{H}^-/G$, given by $\mathbb{H}^-/G^\mu$, where

$$G^\mu = \{f^\mu \circ g \circ (f^\mu)^{-1} : g \in G\}.$$

Also, ψ maps $Q(M) \cap \lambda(T(\mathbb{H}))$ onto $Q(M^\mu) \cap \lambda(T(\mathbb{H}))$.

Now let U be an open ball in $Q(M^\mu)$ centred at the origin and contained in $\lambda_{M^\mu}(T(M^\mu))$. We can write

$$U = Q(M^\mu) \cap \lambda(T(\mathbb{H})) \cap U_0$$

where U_0 is a neighbourhood of the origin in $\lambda(T(\mathbb{H}))$. Since $\lambda(T(\mathbb{H}))$ is open in $Q(\mathbb{H})$, the preimage

$$\psi^{-1}(U) = Q(M) \cap \psi^{-1}(U_0)$$

is a neighbourhood of s_μ in $Q(M)$, and this neighbourhood is contained in

$$\psi^{-1}(\lambda_{M^\mu}(T(M^\mu))) = \lambda_M(T(M)).$$

□

We now show that λ_M is a homeomorphism in a neighbourhood of the origin in $T(M)$.

Theorem 7.6.1. *Let*

$$U = \{[\mu] \in T(M) : d_T([0], [\mu]) < \frac{1}{2} \log 2\}.$$

Then $\lambda_M|_U$ is a homeomorphism onto its image in $Q(M)$.

Proof. The topology on $T(M)$ is the quotient topology from $B(M) \subset L^\infty(M)$ under the equivalence relation $\sim$. This implies that the continuity of λ follows from the continuity of the mapping

$$\mu \mapsto S_{f_\mu|\mathbb{H}}.$$

Now assuming that

$$d_T([0], [\mu_0]) < \frac{1}{2}\log 2,$$

then from the definition of the Teichmüller metric (7.1), $[\mu_0]$ is represented by a Beltrami differential μ_0 for which $||\mu_0||_\infty < 1/3$. In the next section, we see that the mapping $\mu \mapsto S_{f_\mu|\mathbb{H}}$ is holomorphic, and so by the Schwarz lemma,

$$||S_{f_\mu|\mathbb{H}}||_Q \leq 6||\mu||_\infty.$$

Therefore

$$||S_{f_{\mu_0}|\mathbb{H}}||_Q < 2.$$

Let

$$\varphi_0 = S_{f_{\mu_0}|\mathbb{H}},$$

and φ_n be a sequence such that $||\varphi_n - \varphi_0||_Q \to 0$. Then for sufficiently large n, $||\varphi_n|| < 2$, and

$$\lambda_M(-2y^2\varphi_n(\overline{z})) = \varphi_n$$

by Lemma 7.6.1, and also $\mu_0 \sim -2y^2\varphi_0(\overline{z})$. Now $||\varphi_n - \varphi_0|| \to 0$ implies that

$$d_T([-2y^2\varphi_n(\overline{z})], [-2y^2\varphi_0(\overline{z})]) \to 0$$

and so λ_M^{-1} is continuous on $\lambda_M(U)$. □

Exercise 7.6.1. Given any point $[\mu] \in T(M)$, show that λ_M is a homeomorphism in a neighbourhood of $[\mu]$ and thus that λ_M is a homeomorphism.

7.7 Complex structure on Teichmüller space

We can use the Bers embedding to introduce a natural complex analytic structure on Teichmüller space $T(M)$. We first need to define holomorphic functions on Banach spaces. Let X and Y be Banach spaces over $\mathbb{C}$ and let $U \subset X$ be open. A function $f: U \to Y$ has a derivative at a point $x_0 \in U$ if there exists a continuous $\mathbb{C}$-linear mapping $Df(x_0): E \to F$ such that

$$\lim_{h\to 0} \frac{||f(x_0+h) - f(x_0) - Df(x_0)(h)||_Y}{||h||_X} = 0.$$

The mapping $Df(x_0)$ is the derivative of f at x_0. A function $f: U \to Y$ having a derivative at every point of U is called holomorphic in U. A holomorphic function is called biholomorphic if it has a holomorphic inverse.

To provide a connection with usual complex valued analytic functions, we have to introduce the dual space Y^* of Y, which consists if all continuous $\mathbb{C}$-linear mappings of Y into $\mathbb{C}$. The space Y^* can be made into a Banach space by equipping the norm

$$||y^*||_{Y^*} = \sup_{||y||_Y \leq 1} |y^*(y)|.$$

A set $V \subset Y^*$ is called total if $\alpha(y) = 0$ for every $\alpha \in V$ implies that $y = 0$. The following lemma shows how we need only consider complex valued functions of a complex variable when checking whether a mapping between Banach spaces is holomorphic. The proof is given in [31].

Lemma 7.7.1. *A function $f : U \to Y$ is holomorphic if and only if one of the following two conditions are satisfied:*

(a) *For every $x \in U$ and $x_0 \in X$, the function $z \to f(x + zx_0)$ is a holomorphic Y-valued function on every open neighbourhood of $0 \in \mathbb{C}$.*

(b) *The function $f : U \to Y$ is continuous and there exists a total subset V of the dual Y^* such that, for every $\alpha \in V$, the function $\alpha \circ f : U \to \mathbb{C}$ is holomorphic.*

Now, a complex Banach manifold A is a Hausdorff space with an open covering of sets, each of which is homeomorphic to an open subset of a complex Banach space. If A has a maximal atlas for which all parameter transformations are biholomorphic, then A is said to have a complex structure. A manifold A with a complex analytic structure is called a complex Banach manifold.

We have seen that via the Bers embedding, $T(M)$ is globally homeomorphic to an open set in the complex Banach space $Q(M)$. Using this result, it is possible to make Teichmüller space into a complex Banach manifold. However, we will show $T(M)$ has a complex structure in a slightly different way, using local parameter mappings. It is possible to give $T(M)$ a complex structure in several different ways, but it turns out these are all isomorphic, so we can speak of a canonical complex structure on $T(M)$. We will deal here with Teichmüller spaces of Riemann surfaces with $\mathbb{D}$ as their universal cover. For the torus case, if M and N are tori, then there is a quasiconformal map $f : M \to N$. It can be shown (see [31]) that, using the notation of Section 7.4, the bijective isometry $[\varphi] \to [\varphi \circ f^{-1}]$ is actually biholomorphic.

The following result is the foundation for introducing a complex structure on $T(M)$.

Theorem 7.7.1. *The function*

$$\mu \to \Lambda(\mu) = S_{f_\mu|_{\mathbb{H}}}, \tag{7.15}$$

which maps the open unit ball $B(M)$ of $L^\infty(M)$ into the space $Q(M)$, is holomorphic.

Proof. We know that $Q(M)$ is a Banach space, and $B(M)$ is on open subset of the Banach space $L^\infty(M)$. Now fix $\mu, \nu \in B(M)$. For $z \in \mathbb{H}$ and $\varphi \in Q(M)$, define $\alpha_z(\varphi) = \varphi(z)$. Then $V = \{\alpha_z : z \in \mathbb{H}\}$ is a total set in the dual of $Q(M)$. Apply condition (b) of Lemma 7.7.1 to the function

$$w \to s(w) = S_{f_{\mu+w\nu}|_{\mathbb{H}}}. \tag{7.16}$$

The set U in condition (b) is the neighbourhood

$$\left\{ w : |w| < \frac{1 - ||\mu||_\infty}{||\nu||_\infty} \right\}$$

of the origin in $\mathbb{C}$. Then $\mu + w\nu \in B(M)$. Let $Y = Q(G)$ and $\alpha = \alpha_z$. The function

$$w \to \alpha_z \circ s = S_{f_{\mu+w\nu}|_{\mathbb{H}}}(z)$$

is holomorphic in U for every $z \in \mathbb{H}$ by holomorphic dependence on parameters. Also, s is continuous in U, and so by condition (b) of Lemma 7.7.1, the function (7.16) is holomorphic in U. Then from condition (a) of Lemma 7.7.1, the function (7.15) is holomorphic in $B(M)$. □

We now introduce the auxiliary mappings which will be needed to give $T(M)$ a complex structure. For a given $\mu \in B(M)$, write

$$G^\mu = \{f^\mu \circ g \circ (f^\mu)^{-1} : g \in G\},$$

where $M \simeq \mathbb{H}^-/G$, and $M^\mu \simeq \mathbb{H}^-/G^\mu$ is a Riemann surface which is quasiconformally equivalent to M. Recall the mapping $\widetilde{\alpha_\mu} : B(M) \to B(M^\mu)$ given by

$$f^{\widetilde{\alpha_\mu}(\nu)} = f^\nu \circ (f^\mu)^{-1}$$

or, by writing out in full,

$$\widetilde{\alpha_\mu}(\nu) = \left(\frac{\nu - \mu}{1 - \overline{\mu}\nu} \left(\frac{f_z^\mu}{|f_z^\mu|} \right)^2 \right) \circ (f^\mu)^{-1}.$$

The function $\widetilde{\alpha_\mu}$ maps $B(M)$ bijectively onto $B(M^\mu)$, and from Lemma 7.7.1 it follows that $\widetilde{\alpha_\mu}$ is holomorphic. Note that the induced mapping α_μ, given by $\alpha_\mu([\nu]) = [\widetilde{\alpha_\mu}(\nu)]$ is a bijective isometry of $T(M)$ onto $T(M^\mu)$. For $\mu \in B(M)$ and $\nu \in B(M^\mu)$ we write

$$\Lambda_\mu(\nu) = S_{f_\nu|_{\mathbb{H}}}. \tag{7.17}$$

By Theorem 7.7.1, this mapping of $B(M^\mu)$ into $Q(M^\mu)$ is holomorphic. Now, in the ball $B_\mu(0, 1/2) = \{\varphi \in Q(M^\mu) : ||\varphi||_Q < 2\}$, the mapping (7.17) has a section $\sigma_\mu : B_\mu(0, 2) \to B(M^\mu)$ given by

$$\sigma_\mu(\varphi)(\overline{z}) = -2y^2\varphi(z)$$

for $z \in \mathbb{H}$. From Lemma 7.7.1 we see that σ_μ is holomorphic. Let π be the canonical projection of $B(M)$ onto $T(M)$, and λ and λ_μ be the Bers embedding of $T(M)$ and $T(M^\mu)$, respectively, into $Q(M)$ and $Q(M^\mu)$. The collection

$$\{V_\mu = (\pi \circ (\widetilde{\alpha_\mu})^{-1} \circ \sigma_\mu)(B_\mu(0, 1/2)) : \mu \in B(M)\}$$

is an open covering of $T(M)$. Indeed, V_μ is the pre-image of $B_\mu(0,1/2)$ under the homeomorphism $h_\mu = \lambda_\mu \circ \alpha_\mu$ of $T(M)$ onto $T(M^\mu)$.

Theorem 7.7.2. *The atlas*

$$\{(V_\mu, h_\mu) : \mu \in B(M)\} \tag{7.18}$$

defines a complex structure on Teichmüller space $T(M)$. The Bers embedding $[\mu] \to S_{f_\mu|\mathbb{H}}$ of $T(M)$ into $Q(M)$ is holomorphic with respect to this structure.

Proof. Choose two elements $\mu_1, \mu_2 \in B(M)$ such that $V_{\mu_1} \cap V_{\mu_2}$ is non-empty. In $h_{\mu_1}(V_{\mu_1} \cap V_{\mu_2})$, we have

$$h_{\mu_2} \circ h_{\mu_1}^{-1} = \Lambda_{\mu_2} \circ \widetilde{\alpha_{\mu_2}} \circ (\widetilde{\alpha_{\mu_1}})^{-1} \circ \sigma_{\mu_1}.$$

We know that all the mappings on the right-hand side of this equation are holomorphic, and so $h_{\mu_2} \circ h_{\mu_1}^{-1}$ is holomorphic. Switching μ_1 and μ_2 in this calculation shows that $h_{\mu_2} \circ h_{\mu_1}^{-1}$ is biholomorphic, and so (7.18) defines a complex structure for $T(M)$. To show that the Bers embedding is holomorphic, we have to show that $\lambda \circ h_\mu^{-1}$ is holomorphic in $B_\mu(0,1/2)$. Now,

$$\lambda \circ h_\mu^{-1} = \Lambda \circ (\widetilde{\alpha_\mu})^{-1} \circ \sigma_\mu$$

and since all the mappings on the right-hand side are holomorphic, then $\lambda \circ h_\mu^{-1}$ must also be holomorphic. □

Theorem 7.7.3. *The canonical projection*

$$\pi : B(M) \to T(M)$$

is holomorphic and has local holomorphic sections everywhere in $T(M)$.

Proof. From the equation

$$h_\mu \circ \pi = \Lambda_\mu \circ \widetilde{\alpha_\mu}$$

we have that π is holomorphic, since Λ_μ and $\widetilde{\alpha_\mu}$ are. Now consider the mapping

$$\psi_\mu = (\widetilde{\alpha_\mu})^{-1} \circ \sigma_\mu \circ h_\mu$$

of V_μ into $B(M)$. Since all the functions on the right-hand side of this equation are holomorphic, ψ_μ must also be holomorphic. Looking at the definitions, we see that $\pi \circ \psi_\mu$ is the identity mapping on V_μ, and so ψ_μ is the required section. □

In [13], Earle showed that there is no global section of Teichmüller space. The next theorem, proved in [15], shows that holomorphic maps from $\mathbb{D}$ into Teichmüller space can be lifted to holomorphic maps from $\mathbb{D}$ into the space of Beltrami differentials, and uses the idea of holomorphic motions from Chapter 6.

Theorem 7.7.4. *If $f : \mathbb{D} \to T(M)$ is holomorphic, then there exists holomorphic $g : \mathbb{D} \to B(M)$ such that $\pi \circ g = f$. If $\mu_0 \in B(M)$ and $\pi(\mu_0) = f(0)$, then we can choose g such that $g(0) = \mu_0$.*

Proof. By the Bers embedding, we can consider $f(t)$ to be an element of $Q(M)$ for all $t \in \mathbb{D}$. Without loss of generality, we may assume that $f(0) = 0$ (by using right translations, or taking a quasiconformal conjugate). Let $\widehat{\mathbb{R}} = \mathbb{R} \cup \{\infty\}$ and $E = \mathbb{H} \cup \widehat{\mathbb{R}}$. Then E is a G-invariant set, recalling that $M \simeq \mathbb{H}/G$. We shall use f to define a G-equivariant holomorphic motion of E.

For each $t \in \mathbb{D}$, let $\widetilde{f}(t, .)$ be the meromorphic function whose Schwarzian derivative is $f(t)$. Since $f(t) \in T(M)$, $\widetilde{f}(t, .)$ is injective in $\mathbb{H}$ and has a continuous injective extension to E. Therefore, we specify $\widetilde{f}(t, .)$ uniquely by requiring it to fix 0, 1 and ∞. Thus $\widetilde{f}(0, z) = z$ for all $z \in E$. The map $\widetilde{f} : \mathbb{D} \times E \to \widehat{\mathbb{C}}$ is a holomorphic motion of E. Since $f(t)$ is a quadratic differential, for all $t \in \mathbb{D}$ and $g \in G$, there exists a Möbius transformation $\theta_t(g)$ such that

$$\widetilde{f}(t, g(z)) = \theta_t(g)(\widetilde{f}(t, z)),$$

for all $z \in E$. Thus $\widetilde{f}$ is the desired G-equivariant motion. Using the equivariant Slodkowski extension theorem, Theorem 6.2.1, we can extend $\widetilde{f}$ to a G-equivariant holomorphic motion of $\widehat{\mathbb{C}}$, which we still denote by $\widetilde{f}$. For each $t \in \mathbb{D}$, let $g(t)$ be the complex dilatation

$$g(t) = \frac{\widetilde{f}_{\bar{z}}(t, .)}{\widetilde{f}_z(t, .)}$$

in $\mathbb{H}$. The G-equivariance of $\widetilde{f}$ implies that $g(t) \in B(M)$ for all $t \in \mathbb{D}$, and $g : \mathbb{D} \to B(M)$ is holomorphic. By the definition of the Bers embedding, $\pi(g(t))$ is the Schwarzian derivative of $\widetilde{f}(t, .)$ in $\mathbb{H}$, and thus $\pi \circ g = f$.

The lift g satisfies $g(0) = 0$. If μ_0 is a Beltrami differential with $\pi(\mu_0) = 0$, then let $h : B(M) \to B(M)$ be the right translation that sends 0 to μ_0. Then $\pi \circ h = \pi$, and so $h \circ g$ is a lift of f that maps 0 to μ_0. □

Theorem 7.7.5. *The Bers embedding $\lambda : T(M) \to \lambda(T(M)) \subset Q(M)$ is biholomorphic.*

We omit the proof, which can be found in [31]. The following theorem strengthens Theorem 7.3.5, in that we can now show that the isometry between the Teichmüller spaces of quasiconformally equivalent Riemann surfaces can be chosen to be biholomorphic.

Theorem 7.7.6. *Quasiconformally equivalent Riemann surfaces have isometrically and biholomorphically isomorphic Teichmüller spaces.*

Proof. Let $M = \mathbb{H}^-/G$ and $N = \mathbb{H}^-/G'$ be quasiconformally equivalent Riemann surfaces. Consider a lift of a quasiconformal mapping of M onto N to a self-mapping of the lower half-plane $\mathbb{H}^-$. We can assume that the lift is of the form f^μ with $\mu \in B(G)$ and recalling that we can identify $B(G)$ with $B(M)$. The mapping $\widetilde{\alpha_\mu}$ induced by μ is a biholomorphic mapping of $B(G)$ onto $B(G')$, and induces the mapping α_μ of $T(M)$ onto $T(N)$. We have to show that α_μ is biholomorphic. If $\pi' : B(G') \to T(N)$ is the canonical projection, we have

$$\alpha_\mu \circ \pi = \pi' \circ \widetilde{\alpha_\mu}.$$

Now consider an arbitrary element $\nu \in B(G)$. In V_ν, we have

$$\alpha_\mu = \alpha_\mu \circ \pi \circ \psi_\nu = \pi' \circ \widetilde{\alpha_\mu} \circ \psi_\nu.$$

Since all the functions on the right-hand side are holomorphic, so we conclude that α_μ is holomorphic. By switching the roles of M and N, we see that α_μ is in fact biholomorphic, and the theorem is proved. □

Now, if M is a Riemann surface of finite analytic type, that is a compact Riemann surface of genus g with a finite number n of punctures, then $Q(M)$ can be identified with its predual space $A^1(M)$, the subset of $L^1(M)$ consisting of holomorphic quadratic differentials on M with finite norm

$$||\varphi|| = \int_M |\varphi|,$$

for $\varphi \in A^1(M)$.

Exercise 7.7.1. Prove this statement, by showing that every linear functional on $A^1(M)$, $\lambda : A^1(M) \to \mathbb{C}$ has the form

$$\lambda(\varphi) = \int_M \rho_M^{-2} \overline{\psi} \varphi$$

for some $\psi \in Q(M)$, and that $\lambda = 0$ if and only if $\psi = 0$.

The space $A^1(M)$ will be discussed in greater detail in Section 8.2, but for now we will compute the dimension of $A^1(M)$. Let $\widetilde{M}$ be a compact Riemann surface of genus $g \geq 0$ and let E be a finite, possibly empty, subset of $\widetilde{M}$ which contains exactly $n \geq 0$ points. We assume that $2g + n \geq 5$, so that the Riemann surface $M = \widetilde{M} \setminus E$ has non-exceptional finite type.

Each $\varphi \in A^1(M)$ can be regarded as a quadratic differential on $\widetilde{M}$ which is holomorphic except for isolated singularities at the points of M. The integrability of φ implies that the singularities of φ are either removable or simple poles, so $A^1(M)$ is the space of meromorphic quadratic differentials on $\widetilde{M}$ whose poles, if any, are simple and belong to E.

Proposition 7.7.1. *If $M = \widetilde{M} \setminus E$ as above, then the dimension of $A^1(M)$ is $3g - 3 + n \geq 2$. If $x \in E$, then some $\varphi \in A^1(M)$ has a pole at x.*

Proof. A divisor on $\widetilde{M}$ is, by definition, a function $D : \widetilde{M} \to \mathbb{Z}$ such that $D(x) \neq 0$ for only finitely many points $x \in \widetilde{M}$. If D is a divisor on $\widetilde{M}$, its degree is the number

$$\deg(D) = \sum_{x \in \widetilde{M}} D(x).$$

As an example, the characteristic function χ_E of E is a divisor on $\widetilde{M}$ of degree n. If α is a meromorphic function, one-form, or quadratic differential on $\widetilde{M}$ and $x \in \widetilde{M}$, then the order of α at x, $\mathrm{ord}_x(\alpha)$, is 0 if α is holomorphic and non-zero at x, k if α has a zero of order k at x, $-k$ if α has a pole of order k at x and

$+\infty$ if α is identically zero on $\widetilde{M}$. We say α is trivial if it is identically zero on $\widetilde{M}$. If α is non-trivial, then the map $x \mapsto \operatorname{ord}_x(\alpha)$ is a divisor on $\widetilde{M}$, called the divisor of α and denoted by (α).

Let $\mathcal{M}(\widetilde{M})$ be the field of meromorphic functions on $\widetilde{M}$. For any divisor D on $\widetilde{M}$, we define $\mathcal{O}_D(\widetilde{M})$ to be the complex vector space of all functions in $\mathcal{M}(\widetilde{M})$, including the zero function, that are multiples of the divisor $-D$, that is,

$$\mathcal{O}_D(\widetilde{M}) = \{f \in \mathcal{M}(\widetilde{M}) : \operatorname{ord}_x(f) \geq -D(x) \quad \text{for all} \quad x \in \widetilde{M}\}.$$

Now let ω be a non-trivial meromorphic one-form on $\widetilde{M}$, and consider the divisor $D = 2(\omega) + \chi_E$ on $\widetilde{M}$. The space $A^1(M)$ consists precisely of the meromorphic quadratic differentials $\varphi = f\omega^2$ such that $f \in \mathcal{O}_D(\widetilde{M})$.

The divisor of ω has degree $2g - 2$, so $D = 2(\omega) + \chi_E$ has degree $4g - 4 + n$. Since $2g + n \geq 5$ by assumption, we have $\deg(D) \geq 2g + 1$. Consequently, by the Riemann–Roch theorem,

$$\dim(A^1(M)) = \dim(\mathcal{O}_D(\widetilde{M})) = \deg(D) + 1 - g = 3g - 3 + n \geq g + 2 \geq 2.$$

To complete the proof, choose any $x \in E$. Since $\deg(D) \geq 2g + 1$, there exists $f \in \mathcal{O}_D(\widetilde{M})$ with $\operatorname{ord}_x(f) = -D(x)$. By the definition of D, the quadratic differential $f\omega^2 \in A^1(M)$ has a simple pole at x. □

In the case where M is of infinite analytic type, then the dimension of $A^1(M)$ is infinite. For example, if M has an infinite number of punctures at the points (z_n) for $n = 1, 2, \ldots$, then there exist functions f_n, each of which have a simple pole at z_n, and which are linearly independent. This points us in the direction of the following result.

Theorem 7.7.7. *A Riemann surface M is of finite analytic type if and only if the dimension of $A^1(M)$ is finite.*

Exercise 7.7.2. Prove this theorem. Warning: not obvious, see pages 324–335 of [29].

8

Extremal quasiconformal mappings

In this chapter, we investigate extremal quasiconformal mappings, which are extremal in the sense of minimizing the $||.||_\infty$ over all Beltrami differentials corresponding to quasiconformal mappings in the Teichmüller equivalence class. Extremal mappings always exist, but need not be unique, as is shown in several examples. We show how extremality is equivalent to infinitesimal extremality, which is a notion of extremality in a linear vector space, and hence easier to work with than extremality in the Teichmüller equivalence relation. Grötzsch's argument is extended to the Main Inequality which, amongst many applications, can be used to show that Beltrami differentials of the special Teichmüller type are uniquely extremal. See [43] for Reich's survey article on external quasiconformal mappings.

8.1 Examples of extremal mappings

In this chapter, M will be a hyperbolic Riemann surface. We first define what an extremal mapping is.

Definition 8.1.1. *If $[\mu]$ is an element of Teichmüller space $T(M)$, then $\mu_0 \in [\mu]$ is extremal if for any other $\nu \in [\mu]$, we have*

$$||\mu_0||_\infty \leq ||\nu||_\infty. \tag{8.1}$$

It is uniquely extremal if the inequality in (8.1) is strict.

Remark. Via the Beltrami differential equation, we can speak of the Beltrami differential μ or the quasiconformal mapping f^μ being extremal or uniquely extremal.

If f is quasiconformal and M is a hyperbolic Riemann surface, it follows that $f(M)$ is also a hyperbolic Riemann surface. Letting G and G_1 respectively be the covering groups of M and $f(M)$ over $\mathbb{H}$, then f can be lifted to a quasiconformal map $\widetilde{f} : \mathbb{H} \to \mathbb{H}$ such that

$$\gamma_1 \circ \widetilde{f} \circ \gamma = \widetilde{f}, \tag{8.2}$$

for all $\gamma \in G$ and $\gamma_1 \in G_1$.

Finding an extremal quasiconformal mapping in the same Teichmüller class as f is equivalent to finding a map $\widehat{f} : \mathbb{H} \to \mathbb{H}$ which has minimal maximal dilatation over all quasiconformal mappings with the same boundary values, that is, on $\mathbb{R}$, as $\widetilde{f}$ and which satisfy the covering group invariance (8.2). Letting $\mathcal{F}$ be the family of all such mappings, it follows from Theorem 3.4.4 that $\mathcal{F}$ is compact in the C^0-topology. If

$$K_0 = \inf\{K_g : g \in \mathcal{F}\}$$

then take a sequence $g_n \in \mathcal{F}$ such that $K_{g_n} \to K_0$. Since $\mathcal{F}$ is compact, the sequence g_n converges uniformly on compact subsets to a quasiconformal map g. This limit g must have the same boundary values as all the g_n, and must satisfy the group invariance (8.2). Hence $g \in \mathcal{F}$ and $K_g = K_0$. This proves the following theorem.

Theorem 8.1.1. *Every Teichmüller equivalence class $[\mu] \in T(M)$ contains an extremal representative.*

Remark. This extremal representative need not be unique, as we shall soon see.

Now, if f_0 is an extremal mapping in the same class as $\widetilde{f}$, as defined above, let K_0 be the maximal dilatation of f_0, called the extremal dilatation. Clearly, f_0 and K_0 are invariant under conformal transformations, and so we can consider extremal mappings subject to boundary values on a Jordan domain Ω. Since the inverse of a K-quasiconformal mapping is a K-quasiconformal mapping, h^{-1} determines the same value for extremal dilatation K_0, and f_0^{-1} is an extremal mapping for the boundary values h^{-1}.

We saw in Section 1.3 that the Grötzsch argument shows that the affine map is extremal, and indeed uniquely extremal, among all quasiconformal mappings between rectangles which map the four vertices to four corresponding vertices. Now, given a Jordan domain Ω which has finite two-dimensional Lebesgue measure, let h be the restriction of the affine map A_f to $\partial\Omega$. Then the extremal dilatation is again K_{A_f}, and the affine map A_f is uniquely extremal subject to the boundary values h. This can be proved in the same way as for rectangles, by considering the lengths of images of the components of horizontal sections of Ω.

Now let Q and Q' be quadrilaterals in the sense of Section 2.3, that is, they are Jordan domains with distinguished sequential boundary points $\{z_1, z_2, z_3, z_4\}$ and $\{w_1, w_2, w_3, w_4\}$. We can map Q onto Q' quasiconformally by finding conformal maps $\Phi : Q \to R$, $\Psi : Q' \to R'$ onto rectangles R, R' and defining

$$f = \Psi^{-1} \circ A_K \circ \Phi,$$

where $A : R \to R'$ is the affine map whose complex dilatation is K. Using the above arguments, if we ask for a mapping of Q onto Q' so that $z_i \mapsto w_i$ for

$i = 1, \dots, 4$, and is free on the edges between the vertices, then f is the uniquely extremal mapping satisfying these conditions. Letting $k = (K-1)/(K+1)$, the complex dilatation of f turns out to be

$$\mu_f(z) = k\frac{\overline{\varphi(z)}}{|\varphi(z)|} \tag{8.3}$$

for $z \in Q$, and where $\varphi(z) = \Phi'(z)^2$. Now let Q be $\mathbb{D}$ with four distinguished boundary points. Using inequality (1.2), we arrive at the following. Let f be a quasiconformal mapping of $\mathbb{D}$, Q denote $\mathbb{D}$ with four distinguished boundary points and let

$$K_Q = \frac{1-k_Q}{1+k_Q} = \frac{m(f(Q))}{m(Q)}$$

where m is the module defined in Section 2.3. Also let Φ_Q map Q conformally onto a rectangle R of unit area and $\varphi_Q = (\Phi_Q')^2$. Then

$$\frac{k_Q}{1-k_Q} \le \Re \int_{\mathbb{D}} \frac{\mu_f}{1-|\mu_f|^2}\varphi_Q \, dx\, dy + \int_{\mathbb{D}} \frac{|\mu_f|^2}{1-|\mu_f|^2}|\varphi_Q| \, dx\, dy.$$

Varying z along the arcs of ∂Q between the vertices, then $\Phi'(z)dz$ must be real or purely imaginary, depending on whether the arc goes to a horizontal or vertical edge of R. In either case

$$\varphi(z)dz^2 \in \mathbb{R}.$$

Here $\varphi(z)dz^2$ is a quadratic differential which is real on the arcs (z_i, z_{i+1}), and $z_5 = z_1$, of Q. Recall from Chapter 1 the definition of a quadratic differential. For further material on quadratic differentials, we refer the reader to standard references such as [21, 31, 52].

Definition 8.1.2. *A Beltrami differential of the form,*

$$\mu_f(z) = k\frac{\overline{\varphi(z)}}{|\varphi(z)|},$$

as seen in (8.3), where

$$||\varphi||_1 = \int_{\Omega} |\varphi(z)| \, dx\, dy < \infty$$

and $0 < k < 1$ *is called a Beltrami differential of Teichmüller type, and the corresponding quasiconformal mapping* f *is called a Teichmüller mapping. For use in the next chapter, we call* μ_f *a Beltrami differential of generalized Teichmüller type if* $||\varphi||_1 = \infty$.

The uniquely extremal mapping of Q onto Q' is of Teichmüller type, where $||\varphi||_1 = |R|$ and R is the auxiliary rectangle. Considering the f given in Exercise 1.2.2 for $t > 0$, $t \neq 1$, then f is a generalized Teichmüller mapping, since for it, $\varphi(z) = \pm z^{-2}$, which does not belong to $L^1(\mathbb{C})$.

If f is a Teichmüller mapping with $\mu_f(z) = k\overline{\varphi(z)}/|\varphi(z)|$, and $z \leftrightarrow \widetilde{z}$ gives a conformal change of domain between Ω and $\widetilde{\Omega}$, then $f(z)$ is replaced by $\widetilde{f}(\widetilde{z})$, where

$$\mu_{\widetilde{f}}(\widetilde{z}) = \widetilde{k}\frac{\overline{\widetilde{\varphi}(\widetilde{z})}}{|\widetilde{\varphi}(\widetilde{z})|}$$

and where $\widetilde{k} = k$, $\widetilde{\varphi}(\widetilde{z})\mathrm{d}\widetilde{z}^2 = \varphi(z)\mathrm{d}z^2$ and $||\widetilde{\varphi}||_1 = ||\varphi||_1$. Thus $\widetilde{f}$ is also a Teichmüller mapping and the norm is invariant.

Reich and Strebel proved, using the Main Inequality which we will come to, that if f is a Teichmüller mapping of $\mathbb{D}$ with a quadratic differential of finite norm, then f is uniquely extremal for its boundary values. This is Theorem 8.3.3. However, it is possible to have Teichmüller classes which contain no extremal representatives of generalized Teichmüller type. For a construction of such a situation, see [57].

A generalized Teichmüller mapping does not have to be uniquely extremal for its boundary values. For example, consider the affine stretch A of $\mathbb{H}$ given by

$$A(x + iy) = Kx + iy.$$

In this case, the associated quadratic differential does not have finite norm, and the conformal mapping $B(x + iy) = K(x + iy)$ has the same boundary values. We now consider some more examples of extremal mappings.

Example 8.1.1. Let S be the infinite strip

$$\{x + iy : -\infty < x < \infty, -1 < y < 1\},$$

and let A be the affine map given by $A(z) = Kx + iy$. We will show that A is extremal for its boundary values. Let f be any extremal quasiconformal mapping with the same boundary values as A. Let f_1 be the mapping of S given in the following way: subdivide S into two infinite strips $(-\infty, \infty) \times (0, 1)$ and $(-\infty, \infty) \times (-1, 0)$ and let f_1 act on each of these strips as f acts on S by rescaling. That is, on $(-\infty, \infty) \times (0, 1)$,

$$f_1(x + iy) = f(x + i(2y - 1))$$

and on $(-\infty, \infty) \times (-1, 0)$,

$$f_1(x + iy) = f(x + i(2y + 1)).$$

Then f_1 agrees with A on the lines $y = -1$, $y = 0$, and $y = 1$. Define f_n by subdividing S into 2^{-n} strips, with f_n acting on each of these strips as f acts on S. Then we have f_n agreeing with A on each line between the 2^{-n} strips.

Each f_n is an extremal quasiconformal mapping, and each f_n has the same boundary values. By the compactness property of quasiconformal maps, recalling Theorem 3.4.4, f_n converges uniformly on compact subsets to a quasiconformal mapping, which we see must be the affine mapping A. The maximal dilatation of A is not greater than the maximal dilatation of each f_n, since $f_n \to A$ uniformly on compact subsets and furthermore, since each f_n is extremal, A must also be extremal. In fact, Strebel proved that A is uniquely extremal, which we will show later.

Example 8.1.2. The finiteness of the norm of a Teichmüller mapping is not necessary for unique extremality of Teichmüller mappings. Recall from Exercise 1.2.2 the function $w = f(z) = r^z e^{i\theta}$. The associated quadratic differential is $\varphi(z) = 1/z^2$. Since

$$\int_{\mathbb{H}} \left| \frac{1}{z^2} \right| \, dx \, dy = \infty,$$

then f is a generalized Teichmüller mapping of $\mathbb{H}$ onto $\mathbb{H}$ with infinite norm. Now, replacing z and w with $\log z$ and $\log w$ respectively, f is transformed to a horizontal stretch between strips, which we will show is uniquely extremal.

Example 8.1.3. Strebel's Chimney [49].

This was the first example of a quasiconformal map that is extremal for its boundary values, but not uniquely extremal. Let Ω be the lower half-plane $\mathbb{H}^-$ together with the half-strip $(0, 1) \times [0, \infty)$.

Strebel proved that the affine stretching $A(x + iy) = Kx + iy$ is extremal for the boundary values it induces. However, the quasiconformal mapping

$$f_1(z) = \begin{cases} A(z), & 0 < x < 1, y \geq 0, \\ Kz, & y < 0, \end{cases}$$

has the same boundary values as A, and $K_{f_1} = K_A = K$. Thus f_1 is also extremal, but different from A.

Example 8.1.4. [49] Let Ω_α be the angular region

$$\Omega_\alpha = \left\{ z : -\frac{\alpha}{2} < \arg z < \frac{\alpha}{2} \right\}$$

for $0 < \alpha < 2\pi$. It turns out that the affine stretch A is not extremal. In the case $\alpha = \pi$, this is clear as the identity mapping has the same boundary values as A. For $\alpha \neq \pi$, the extremal mapping can actually be computed. Let $\zeta = \log z$ and

$\lambda = \log w$. The strip

$$S = \{\zeta : |\mathrm{Im}(\zeta)| < \alpha/2\}$$

is mapped onto the strip

$$S' = \{\lambda : |\mathrm{Im}(\lambda)| < \beta/2\},$$

where

$$\beta = 2\tan^{-1}\left(\frac{1}{K}\tan\frac{\alpha}{2}\right).$$

The boundary values $\zeta \to \lambda$ correspond to a shift to the right in the amount of

$$s = \log\sqrt{K^2\cos^2\left(\frac{\alpha}{2}\right) + \sin^2\left(\frac{\alpha}{2}\right)}.$$

We know from previous examples that there is a uniquely extremal mapping from S onto S', which is an affine mapping shifting to the right by a factor s, and a compression by the factor α/β. Now returning to $z \to w$, we can conclude there is a uniquely extremal Teichmüller mapping, with infinite norm in this case, and

$$K_0 = \frac{\alpha}{\beta} = \frac{\alpha}{2\tan^{-1}(\frac{1}{K}\tan(\alpha/2))} < K.$$

As we can see from these examples, the $[\mu]$ equivalence relation is very complicated and highly non-linear, which makes finding extremal quasiconformal maps very difficult. However, there is another equivalence relation $[\mu]_*$, which we shall explore in the coming sections. We shall see that this equivalence relation is linear, and it turns out that extremality in one relation is equivalent to extremality in the other, which gives us new ways to explore extremal quasiconformal mappings.

8.2 The Hamilton–Krushkal condition

The Hamilton–Krushkal condition gives a necessary and sufficient condition for a complex dilatation to be extremal. In this section we will prove the necessity, and leave the sufficiency until later. Now let M be a Riemann surface with covering group G over $\mathbb{H}$. Recall that $L^\infty(M)$ is the Banach space of bounded measurable functions on M. Note that we can identify $L^\infty(M)$ with $L^\infty(G)$, which is the Banach space of bounded measurable functions μ on $\mathbb{H}$ which satisfy

$$\frac{\mu(A(z))\overline{A'(z)}}{A'(z)} = \mu(z)$$

for every $A \in G$ and $z \in \mathbb{H}$. We define $B(M)$ to be the unit ball of $L^\infty(M)$. The space $A^1(M)$ consists of holomorphic quadratic differentials φ on M which have finite norm

$$||\varphi||_1 = \int_M |\varphi| \,\mathrm{d}x\,\mathrm{d}y < \infty.$$

The space $A^1(M)$ is called the Bergman space of M. The analogous space $A^1(G)$, which can be identified with $A^1(M)$, consists of holomorphic quadratic differentials on $\mathbb{H}$ which satisfy

$$\varphi(A(z))A'(z)^2 = \varphi(z)$$

for all $A \in G$ and $z \in \mathbb{H}$, and finite 1-norm where the region of integration is a fundamental region of G.

Exercise 8.2.1. Prove that $A^1(M)$ is a Banach space.

For $\mu \in L^\infty(M)$, consider the linear functional

$$\Lambda_\mu(\varphi) = \int_M \mu(z)\varphi(z) \,\mathrm{d}x\,\mathrm{d}y \tag{8.4}$$

for $\varphi \in A^1(M)$. The norm of Λ_μ is given by

$$||\Lambda_\mu|| = \sup\{|\Lambda_\mu(\varphi)| : \varphi \in A^1(G), ||\varphi||_1 = 1\}. \tag{8.5}$$

Here we introduce the equivalence relation promised at the end of the previous section.

Definition 8.2.1. *For $\mu, \nu \in L^\infty(M)$, we say μ and ν are infinitesimally equivalent, or $\mu \sim_* \nu$, if*

$$\Lambda_\mu(\varphi) = \Lambda_\nu(\varphi)$$

for all $\varphi \in A^1(M)$. The infinitesimal equivalence class of μ is denoted by $[\mu]_*$.

The space of infinitesimally trivial Beltrami differentials for M is given by

$$N(M) = \left\{\mu \in L^\infty(M) : \int_M \mu\varphi = 0, \text{ for all } \varphi \in A^1(M)\right\}.$$

So for any $\mu \in L^\infty(M)$ and $\nu \in N(M)$, we have $\mu + \nu \sim_* \mu$. Also, $B_0(M)$ denotes the space of Beltrami coefficients which are equivalent to 0 under the Teichmüller equivalence relation. The notion of extremality also applies to the infinitesimal equivalence relation.

Definition 8.2.2. *An element μ of $L^\infty(M)$ is said to be infinitesimally extremal if for all $\nu \in [\mu]_*$,*

$$||\mu||_\infty \leq ||\nu||_\infty,$$

and μ is called infinitesimally uniquely extremal if the inequality is strict.

Now Λ_μ is a linear functional on $A^1(M)$, and via the Hahn–Banach theorem (see [47]) can be extended to a linear functional $\widetilde{\Lambda_\mu} : L^1(M) \to \mathbb{C}$ which agrees with Λ_μ on $A^1(M)$. Since the Banach dual of $L^1(M)$ is $L^\infty(M)$, then by the Riesz representation theorem (again, see [47]), there exists $\nu \in L^\infty(M)$ such that

$$\widetilde{\Lambda_\mu}(\varphi) = \int_M \nu\varphi,$$

for all $\varphi \in L^1(M)$. Since $\widetilde{\Lambda_\mu}$ agrees with Λ_μ on $A^1(M)$, we have that $\nu \in [\mu]_*$, and so $\Lambda_\mu = \Lambda_\nu$ as linear functionals on $A^1(M)$. Since also the Hahn–Banach extension is norm-preserving, by choosing an appropriate element of $L^1(M)$, $||\widetilde{\Lambda_\mu}|| = ||\nu||_\infty$. In particular, this shows that ν is an infinitesimally extremal element of $[\mu]_*$ and so infinitesimal extremals exist. Note that the infinitesimal extremal need not be unique since it depends on the choice implicit in the Hahn–Banach extension. Also this shows that ν is infinitesimally extremal if and only if $||\Lambda_\nu|| = ||\nu||_\infty$.

In the case where $A^1(M)$ is finite dimensional and ν is extremal, Λ_ν attains its norm for some $\varphi_0 \in A^1(M)$. That is,

$$\int_M \nu\varphi_0 = ||\nu||_\infty$$

and the only possibility is that $\nu = \overline{\varphi_0}/|\varphi_0|$. This shows that, in this case, ν is uniquely infinitesimally extremal and of Teichmüller type.

The Hamilton–Krushkal condition is the statement that μ is extremal in its Teichmüller class if and only if it is infinitesimally extremal. Before we prove the necessity of the Hamilton–Krushkal condition, we need a variational lemma, the proof of which can be found in [21].

Lemma 8.2.1. *A Beltrami differential μ is in the space $N(M)$ if and only if there exists a real analytic curve ν_t of Beltrami coefficients in $B_0(M)$, for which*

$$\nu_t(z) = t\mu(z) + O(t^2)$$

uniformly in $z \in M$.

Theorem 8.2.1. *If μ is an extremal Beltrami coefficient, then*

$$||\Lambda_\mu|| = ||\mu||_\infty,$$

and μ is infinitesimally extremal.

Proof. Assume that $k = ||\mu||_\infty > ||\Lambda_\mu||$. By the Hahn–Banach extension theorem and the Riesz representation theorem, there exists $\tau \in B(G)$ such that

$$\int_\Omega \mu\varphi \, dx \, dy = \int_\Omega \tau\varphi \, dx \, dy$$

for all $\varphi \in A^1(G)$, and such that $||\tau||_\infty = ||\Lambda_\mu||$. Then $\mu - \tau$ is infinitesimally trivial, and by Lemma 8.2.1, there exists a curve of trivial Beltrami differentials ν_t such that

$$||\nu_t - t(\mu - \tau)||_\infty = O(t^2). \tag{8.6}$$

Letting $\nu = \nu_t$ for brevity, form

$$f^\sigma = f^\mu \circ (f^\nu)^{-1}.$$

It is clear that σ has the same Teichmüller class as μ. For small enough $t > 0$, we will show that $||\sigma||_\infty < ||\mu||_\infty$, which would contradict the assumption on μ being extremal. Using (1.5),

$$\sigma(f^\nu(z)) = \frac{\mu - \nu}{1 - \overline{\nu}\mu} \frac{f_z^\nu}{\overline{f_z^\nu}}. \tag{8.7}$$

Squaring this gives

$$|\sigma \circ f^\nu|^2 = \frac{|\mu|^2 - 2\Re(\mu\overline{\nu}) + |\nu|^2}{1 - 2\Re(\mu\overline{\nu}) + |\nu\mu|^2}.$$

Substituting (8.6) into this gives

$$|\sigma \circ f^\nu| = |\mu| - t\frac{1 - |\mu|^2}{|\mu|}\Re(|\mu|^2 - \mu\overline{\tau}) + O(t^2). \tag{8.8}$$

We can use this expression to give useful estimates on $|\sigma \circ f^\nu|$. Recall that $||\Lambda_\mu|| = ||\nu||_\infty < k = ||\mu||_\infty$, and let

$$\Omega_1 = \left\{ z \in \mathbb{H} : |\mu(z)| \leq \frac{(||\Lambda_\mu|| + k)}{2} \right\},$$

$$\Omega_2 = \left\{ z \in \mathbb{H} : \frac{(||\Lambda_\mu|| + k)}{2} < |\mu(z)| \leq k \right\}.$$

We have $\Omega_1 \cup \Omega_2 = \mathbb{H}$, and using (8.8) there exists $\delta_1 > 0$ and $C_1 > 0$ such that for $0 < t < \delta_1$,

$$|\sigma \circ f^\nu(z)| \leq k - C_1 t$$

for $z \in \Omega_1$. For $z \in \Omega_2$, the coefficient of t in (8.8) is bounded below by

$$\frac{1 - k^2}{k}\left(\left(\frac{||\Lambda_\mu|| + k}{2}\right)^2 - ||\Lambda_\mu||k\right) = \frac{(1 - k^2)(k - ||\Lambda_\mu||)^2}{4k} > 0.$$

Thus (8.8) implies there exists $\delta_2 > 0$ and $C_2 > 0$ such that for $0 < t < \delta_2$,

$$|\sigma \circ f^\nu(z)| \leq k - C_2 t$$

for $z \in \Omega_2$. Thus $||\sigma||_\infty < k$ for small enough $t > 0$, which completes the proof. □

Definition 8.2.3. *A sequence $\varphi_n \in A^1(M)$ is called a Hamilton sequence for μ if $||\varphi_n||_1 = 1$ for all n, φ_n converges locally uniformly on M, and*

$$\lim_{n \to \infty} \Lambda_\mu(\varphi_n) = ||\Lambda_\mu||.$$

A Hamilton sequence is called degenerating if it converges locally uniformly to zero.

The fact that a Hamilton sequence exists follows from the following lemma.

Lemma 8.2.2. *The closed unit ball in $A^1(M)$ is a normal family.*

Proof. For a plane domain Ω, the Cauchy integral formula gives

$$|\varphi(z_0)| \le \frac{1}{\pi t^2}\int_0^{2\pi}\int_0^t |\varphi(z_0 + re^{i\theta})| r \, \mathrm{d}r \, \mathrm{d}\theta,$$

for any disk centred at z_0 of radius t which is completely contained in Ω. Thus if $\varphi \in A^1(\Omega)$ with $||\varphi|| < 1$, we have

$$|\varphi(z_0)| \le \frac{1}{\pi t^2}.$$

In particular, if K is any compact subset of Ω, there exists a constant $C(K)$ depending only on K such that

$$|\varphi(z)| \le C(K)$$

for any $\varphi \in A^1(\Omega)$ with $||\varphi|| \le 1$ and $z \in K$. This argument lifts to M via the covering map, by taking Ω to be a fundamental region for M. Therefore the family $\{\varphi|_K : ||f\varphi|| \le 1\}$, where K is a compact subset of M, is uniformly bounded. Using the Arzela–Ascoli theorem, this shows that the closed unit ball of $A^1(M)$ is a normal family, that is, any sequence φ_n in the closed unit ball of $A^1(M)$ has a subsequence which converges locally uniformly to some quadratic differential φ in the closed unit ball of $A^1(M)$. □

We can therefore assume that any Hamilton sequence has a limit quadratic differential $\varphi_0 \in A^1(M)$. If μ is infinitesimally extremal, and if Hamilton sequence is degenerating, it follows that μ can change on a compact subset of M to a different Beltrami differential ν and the same Hamilton sequence will still work for ν. Hence ν will also be infinitesimally extremal.

Example 8.2.1. We show how Theorem 8.2.1 can be applied by looking at the affine stretch of the angular region Ω_α given in Example 8.1.4, and giving an alternative proof. Since the complex dilatation of the affine stretch is constant, it follows from Theorem 8.2.1 that for the affine stretch of a simply connected region R to be extremal, we need

$$||\Lambda_1|| = \sup\left\{\frac{|\int_R \varphi(z)\,\mathrm{d}x\,\mathrm{d}y|}{\int_R |\varphi(z)|\,\mathrm{d}x\,\mathrm{d}y} : \varphi \in A^1(R)\right\} = 1.$$

Let $S = \{w = u + iv : -\alpha/2 < v < \alpha/2\}$. Then

$$I = \int_{\Omega_\alpha} \varphi(z)\,\mathrm{d}x\,\mathrm{d}y = \int_S \mathrm{e}^{-2iv} f(w)\,\mathrm{d}u\,\mathrm{d}v$$

where $f(w) = e^{2w}\varphi(e^w)$, and

$$J = \int_{\Omega_\alpha} |\varphi(z)|\,dx\,dy = \int_S |f(w)|\,du\,dv$$
$$= \int_{-\alpha/2}^{\alpha/2} dv \int_{-\infty}^{\infty} |f(u+iv)|\,dv.$$

Since $J < \infty$, then $\int_{-\infty}^{\infty} f(u+iv)\,du$ exists for almost all v, and since $f(w)$ is holomorphic in S, then

$$\int_{-\infty}^{\infty} f(u+iv)\,du = C$$

for almost all v, where C is a constant. Therefore

$$I = C\int_{-\alpha/2}^{\alpha/2} e^{-2iv}\,dv = C\sin\alpha$$

and $\alpha|C| \le J$. Hence $||\Lambda_1|| \le |\sin\alpha|/\alpha < 1$, and again we see that the affine stretch of Ω_α is not extremal.

Example 8.2.2. Suppose that μ is an extremal Beltrami coefficient defined in $\mathbb{D}$, $k = ||\mu||_\infty > 0$ and for some $0 < \rho < 1$,

$$\mu(z) = s(z)\frac{\overline{\varphi(z)}}{|\varphi(z)|}$$

for $z \in A = \{z : \rho \le |z| < 1\}$, where $s(z)$ is a non-negative measurable function in A, $\varphi(z)$ is holomorphic in A, $|\varphi(z)| \ge m > 0$ for $z \in A$, and $J = \int_A |\varphi| < \infty$. Then φ has a holomorphic extension to all of $\mathbb{D}$, $s(z) \equiv k$ and

$$\mu(z) = k\frac{\overline{\varphi(z)}}{|\varphi(z)|}$$

for $z \in \mathbb{D}$.

Proof. Let $S = \{z \in A : |\mu(z)| \le k/2\}$. By Theorem 8.2.1, there is a Hamilton sequence φ_n such that

$$\delta_n = \int_{\mathbb{D}} (k|\varphi_n| - \Re(\mu\varphi_n))\,dx\,dy = o(1)$$

as $n \to \infty$. Thus

$$\int_S |\varphi_n|\,dx\,dy = o(1)$$

and, since $|\varphi(z)| \ge m > 0$ in A,

$$\int_S \left|\frac{\varphi_n}{\varphi}\right|\,dx\,dy = o(1).$$

If the Hamilton sequence converged uniformly on compact subsets to zero, then

$$\begin{aligned}\delta_n &= \int_{A\setminus S} (k|\varphi_n| - \Re(\mu\varphi_n))\, dx\, dy + o(1)\\ &= \int_{A\setminus S} \left(\frac{k}{s}\left|\frac{\varphi_n}{\varphi}\right| - \Re\left(\frac{\varphi_n}{\varphi}\right)\right) s|\varphi|\, dx\, dy + o(1).\end{aligned}$$

Since the integrand is non-negative,

$$\delta_n \geq \frac{km}{2}\int_{A\setminus S} \left(\frac{k}{s}\left|\frac{\varphi_n}{\varphi}\right| - \Re\left(\frac{\varphi_n}{\varphi}\right)\right) dx\, dy + o(1),$$

and since we are assuming that φ_n degenerates,

$$\int_A \frac{\varphi_n}{\varphi}\, dx\, dy = \frac{1-\rho^2}{2}\int_0^{2\pi} \frac{\varphi_n(\rho e^{i\theta})}{\varphi(\rho e^{i\theta})}\, d\theta = o(1).$$

By Schwarz's inequality,

$$\int_A \left|\frac{\varphi_n}{\varphi}\right| dx\, dy \geq \frac{1}{J}\left(\int_A |\varphi_n|\, dx\, dy\right)^2 = \frac{1+o(1)}{J}.$$

We can therefore conclude that

$$\begin{aligned}\delta_n &\geq \frac{km}{2}\int_{A\setminus S} \frac{k}{s}\left|\frac{\varphi_n}{\varphi}\right| dx\, dy + o(1)\\ &\geq \frac{km}{2}\int_{A\setminus S} \left|\frac{\varphi_n}{\varphi}\right| dx\, dy + o(1)\\ &\geq \frac{km}{2J} + o(1),\end{aligned}$$

which is a contradiction, since we know that $\delta_n = o(1)$ as $n \to \infty$. We conclude that μ must be of Teichmüller type. □

Definition 8.2.4. *If $\mu \in L^\infty(G)$, then $\nu \in L^\infty(G)$ is called an admissible variation of μ if $||\mu||_\infty = ||\nu||_\infty$, and if for some set $S \subset \Omega$, including the possibility that $S = \emptyset$, and for some $s \in \mathbb{R}$, $|\mu(z)| < s < ||\mu||_\infty$ in S and $\nu(z) = \mu(z)$ almost everywhere in $\Omega \setminus S$. The definition only makes sense if $\Omega \setminus S$ has positive measure. In the case that $|\mu(z)|$ is constant almost everywhere, then μ is the only admissible variation of μ.*

The notion of an admissible variation was introduced in [9], and will be important later for characterizing unique extremality.

Theorem 8.2.2. *Let μ_g be an admissible variation of μ_f. Then if f is extremal, so is g.*

Proof. Using Theorem 8.2.1, we have

$$||\Lambda_{\mu_f}|| = ||\mu_f||_\infty = ||\mu_g||_\infty. \tag{8.9}$$

Suppose that $||\Lambda_{\mu_g}|| < ||\mu_g||_\infty$. Then there exists $\widetilde{\eta} \in L^\infty(G)$ such that $\widetilde{\eta} - \mu_g$ is infinitesimally trivial, and

$$||\widetilde{\eta}||_\infty = ||\mu_g||_\infty - \delta = ||\mu_f||_\infty - \delta$$

for some $\delta > 0$. Let

$$\widetilde{\mu}(z) = \mu_f(z) + \frac{||\mu_f||_\infty - s}{2||\mu_f||_\infty}[\widetilde{\eta}(z) - \mu_g(z)],$$

so that $\widetilde{\mu} - \mu_f$ is infinitesimally trivial. When $z \in S$, $|\mu_f| < s$, and so

$$|\widetilde{\mu}(z)| < s + \frac{||\mu_f||_\infty - s}{2||\mu_f||_\infty}[2||\mu_f||_\infty - \delta] = ||\mu_f||_\infty - \frac{\delta(||\mu_f||_\infty - s)}{2||\mu_f||_\infty}.$$

Now when $z \in \Omega \setminus S$, then $\mu_g(z) = \mu_f(z)$, and therefore

$$\widetilde{\mu}(z) = \left(\frac{1}{2} - \frac{s}{2||\mu_f||_\infty}\right)\widetilde{\eta}(z) + \left(\frac{1}{2} + \frac{s}{2||\mu_f||_\infty}\right)\mu_f(z),$$

and thus

$$|\widetilde{\mu}(z)| \leq \left(\frac{1}{2} - \frac{s}{2||\mu_f||_\infty}\right)(||\mu_f||_\infty - \delta) + \left(\frac{1}{2} + \frac{s}{2||\mu_f||_\infty}\right)||\mu_f||_\infty.$$

Hence $||\widetilde{\mu}||_\infty < ||\mu_f||_\infty$, which contradicts (8.9). Therefore $||\Lambda_{\mu_g}|| = ||\mu_g||_\infty$, and so g is extremal. □

8.3 The Main Inequality

In this section we introduce the main tool used in the theory of extremal mappings, the Main Inequality. First, the length-area inequality (1.2) used in the Grötszch argument can be generalized to the Reich–Strebel inequality.

Theorem 8.3.1. *Let M be a Riemann surface, and φ a holomorphic quadratic differential with $||\varphi||_1 < \infty$. Let f be a quasiconformal self-mapping of M homotopic to the identity, and let μ be the Beltrami coefficient of f. Then*

$$||\varphi||_1 \leq \int_M |\varphi(z)| \frac{\left|1 + \mu(z)\frac{\varphi(z)}{|\varphi(z)|}\right|^2}{1 - |\mu(z)|^2}\, dx\, dy. \tag{8.10}$$

We omit the proof, which uses the theory of trajectories of quadratic differentials, which can be found in [22].

The Reich–Strebel inequality can be used to compare the Beltrami coefficient μ of a quasiconformal map f from M onto $f(M)$ with the Beltrami coefficient $\widetilde{\mu}$ of a map $\widetilde{f}$ from $f(M)$ onto M with the property that $\widetilde{f} \circ f$ is homotopic to the identity. If φ is a holomorphic quadratic differential on M normalized so that $||\varphi||_1 = 1$, then (8.10) leads to the Main Inequality (8.11).

Theorem 8.3.2. *Using the notation above, let*

$$\mu_1 = \frac{\widetilde{f}_{\overline{w}}}{\widetilde{f}_w}$$

and

$$\theta = \frac{\overline{p}}{p}\left(1 - \frac{\overline{\mu\varphi}}{|\varphi|}\right)\left(1 - \frac{\mu\varphi}{|\varphi|}\right)^{-1}$$

where $p = f_{\overline{z}}/f_z$. Then

$$1 \leq \int_M \frac{\left|1 + \mu\frac{\varphi}{|\varphi|}\right|^2}{1 - |\mu|^2} \cdot \frac{\left|1 + \theta\mu_1(f)\frac{\varphi}{|\varphi|}\right|^2}{1 - |\mu_1(f)|^2} |\varphi| \, dx \, dy. \tag{8.11}$$

The Main Inequality was first obtained in the form (8.11) for the unit disk by Reich and Strebel in [44], and extended to arbitrary domains by Strebel in [51]. The proof follows from applying Theorem 8.3.1 to the map $(\widetilde{f} \circ f)$. This is a calculation which we shall omit, but is contained in [21]. The Main Inequality can be used to prove the result that if a Beltrami differential is of Teichmüller type, then it is uniquely extremal. This is the proof promised in Section 8.1.

Theorem 8.3.3. *Let $f : M \to N$ be a map between Riemann surfaces M and N which has Beltrami differential of Teichmüller type, that is, $\mu_f = k\overline{\varphi}/|\varphi|$, where $0 \leq k < 1$ and $\varphi \in A^1(M)$. Let $g : M \to N$ be a quasiconformal map such that $g^{-1} \circ f$ is Teichmüller trivial, and let μ_g be the Beltrami differential for g. Then either there exists a set of positive measure on M for which $|\mu_g(z)| > k$ or alternatively $\mu_g(z) = k\overline{\varphi(z)}/|\varphi(z)|$ almost everywhere.*

Proof. Since f and g are in the same Teichmüller class, then $g^{-1} \circ f$ is Teichmüller trivial. Thus we can apply the Main Inequality (8.11) to $g^{-1} \circ f$, which gives

$$1 \leq \int_M |\varphi(z)| \frac{|1 - \mu_f\varphi/|\varphi||^2}{1 - |\mu_f|^2} \cdot \frac{|1 + \mu_g\theta\varphi/|\varphi||^2}{1 - |\mu_g|^2} \, dx \, dy, \tag{8.12}$$

where

$$\theta = \left(1 - \frac{\overline{\mu_f\varphi}}{|\varphi|}\right)\left(1 - \frac{\mu_f\varphi}{|\varphi|}\right)^{-1}$$

and $||\varphi||_1 = 1$. Since μ_f is of Teichmüller type, then $\theta = 1$ and

$$\left(\left|1 - \frac{\mu_f\varphi}{|\varphi|}\right|^2\right)(1 - |\mu_f|^2)^{-1} = K^{-1},$$

where $K = (1 + k)(1 - k)^{-1}$. Therefore (8.12) becomes

$$K \leq \int_M |\varphi| \frac{|1 + \mu_g\varphi/|\varphi||^2}{1 - |\mu_g|^2} \, dx \, dy \tag{8.13}$$

for every μ_g equivalent to $\mu_f = k\overline{\varphi}/|\varphi|$. Now if $|\mu_g(z)| \leq k$, then

$$\frac{1+k}{1-k} = K \leq \int_M \frac{|1+\mu_g\varphi/|\varphi||^2}{1-|\mu_g|^2}|\varphi|\,dx\,dy$$
$$\leq \int_M \frac{1+|\mu_g(z)|}{1-|\mu_g(z)|}|\varphi|\,dx\,dy \leq \frac{1+||\mu_g||_\infty}{1-||\mu_g||_\infty} \leq \frac{1-k}{1+k}.$$

Therefore each inequality in this string of inequalities must be equality, which implies that

$$\frac{|1+\mu_g\varphi/|\varphi||^2}{1-|\mu_g|^2} = \frac{1+k}{1-k}$$

almost everywhere. Thus $\mu_g = k\overline{\varphi}/|\varphi| = \mu_f$ almost everywhere. □

8.4 Sufficiency of the Hamilton–Krushkal condition

To prove the converse to Theorem 8.2.1, we start by considering the following functionals:

$$I(\mu) = \sup\left|\Re\int_M \frac{\mu\varphi}{1-|\mu|^2}\,dx\,dy\right|, \tag{8.14}$$

and

$$J(\mu) = \sup\int_M \frac{|\mu|^2|\varphi|}{1-|\mu|^2}\,dx\,dy, \tag{8.15}$$

where the supremum in both cases is taken over all $\varphi \in A^1(M)$ with $||\varphi||_1 = 1$, and μ is a Beltrami differential. If ν is extremal in the same Teichmüller class as μ, with $||\nu||_\infty = k_0$, and writing $K_0 = (1+k_0)(1-k_0)^{-1}$, then the Main Inequality (8.11) becomes

$$\frac{1}{K_0} \leq \int_M |\varphi|\frac{|1-\mu\varphi/|\varphi||^2}{1-|\mu|^2}\,dx\,dy.$$

Expanding the numerator and simplifying gives

$$\frac{1}{K_0} - 1 \leq -2\Re\int_M \frac{\mu\varphi}{1-|\mu|^2}\,dx\,dy + 2\int_M \frac{|\mu|^2|\varphi|}{1-|\mu|^2}\,dx\,dy$$

which in turn leads to

$$\Re\int_M \frac{\mu\varphi}{1-|\mu|^2}\,dx\,dy \leq \frac{k_0}{1+k_0} + \int_M \frac{|\mu|^2|\varphi|}{1-|\mu|^2}\,dx\,dy,$$

which can be written as

$$I(\mu) \leq \frac{k_0}{1+k_0} + J(\mu). \tag{8.16}$$

An inequality in the opposite direction follows from an application of the Main Inequality (8.11) where ν is extremal in the Teichmüller class of μ. If $||\nu||_\infty = k_0$ and $K_0 = (1+k_0)(1-k_0)^{-1}$, then

$$K_0 \leq \sup \int_M |\varphi| \frac{|1+\mu\varphi/|\varphi||^2}{1-|\mu|^2}\, dx\, dy,$$

where the supremum is taken over all $\varphi \in A^1(M)$ of norm 1. Expanding the numerator and simplifying leads to

$$\frac{k_0}{1-k_0} \leq I(\mu) + J(\mu). \tag{8.17}$$

Combining (8.16) and (8.17) gives the following theorem.

Theorem 8.4.1. *Let $\mu, \nu \in B(M)$ where ν is extremal in the Teichmüller class of μ, and let $k_0 = ||\nu||_\infty$. Then*

$$I(\mu) - \delta(\mu) \leq \frac{k_0}{1-k_0^2} \leq I(\mu) + \delta(\mu), \tag{8.18}$$

where $\delta(\mu) = J(\mu) - k_0^2(1-k_0^2)^{-1}$.

Theorem 8.4.2. *A necessary and sufficient condition for $\mu \in B(M)$ to be extremal in its Teichmüller class is that*

$$I(\mu) = \frac{k}{1-k^2}, \tag{8.19}$$

where $||\mu||_\infty = k$, and the supremum for $I(\mu)$ is over all $\varphi \in A^1(M)$ such that $||\varphi||_1 = 1$.

Proof. Suppose that (8.19) holds. Then the left-hand side of (8.18) gives

$$\frac{k}{1-k^2} \leq \frac{k_0}{1-k_0^2} + \frac{k^2}{1-k^2} - \frac{k_0^2}{1-k_0^2},$$

which implies that

$$\frac{k}{1+k} \leq \frac{k_0}{1+k_0}$$

and so that $k \leq k_0$. Conversely, suppose that μ is extremal, so that $||\mu||_\infty = k_0$ and the right-hand side of (8.18) gives

$$\frac{k_0}{1-k_0^2} \leq I(\mu).$$

The opposite inequality clearly holds, and so we obtain (8.19), which proves the theorem. □

Lemma 8.4.1. *Suppose that φ_n is a Hamilton sequence for μ, that is $||\varphi_n||_1 = 1$ for all n, and*

$$\int_M \varphi_n \mu\, dx\, dy \to k = ||\mu||_\infty$$

as $n \to \infty$. *Then* φ_n *converges uniformly to zero on any set of the form*

$$E = \{z \in M : |\mu(z)| \leq k' < k\}.$$

Proof. Write

$$\int_M \varphi_n \mu = \int_{M \setminus E} \varphi_n \mu + \int_E \varphi_n \mu.$$

The second integral on the right-hand side is bounded above by $k'||\varphi_n||_E$, where $||.||_E$ here denotes the L^1 norm on E. Then

$$\left| \int_M \varphi_n \mu \right| \leq k'||\varphi_n||_E + k||\varphi_n||_{M\setminus E} = (k' - k)||\varphi_n||_E + k.$$

Since the left-hand side of this inequality converges to k as $n \to \infty$, and since $k' < k$, then we must have that $||\varphi_n||_E \to 0$. □

Lemma 8.4.2. *Suppose* φ_n *is a Hamilton sequence for* μ, *and* $||\mu||_\infty = k$. *Then*

$$\int_M \frac{\varphi_n \mu}{1 - |\mu|^2}\, \mathrm{d}x\, \mathrm{d}y \to \frac{k}{1 - k^2}.$$

Proof. Let $k' < k$ and $E = \{z \in M : |\mu| \leq k'\}$. From Lemma 8.4.1, we know that $||\varphi_n||_E \to 0$. Therefore

$$\int_{M\setminus E} \varphi_n \mu \to k$$

and

$$\int_{M\setminus E} \frac{\varphi_n \mu}{1 - k'^2} \to \frac{k}{1 - k'^2}.$$

Since $|\mu| \geq k'$ on $M \setminus E$, then

$$\int_{M\setminus E} \frac{\varphi_n}{1 - |\mu|^2}$$

is a sequence whose cluster points are between $k(1 - k'^2)^{-1}$ and $k(1 - k^2)^{-1}$. Since $||\varphi_n||_E \to 0$, this implies that

$$\int_M \frac{\varphi_n \mu}{1 - |\mu|^2}$$

is a sequence whose cluster points are between $k(1 - k'^2)^{-1}$ and $k(1 - k^2)^{-1}$. Since k' is an arbitrary number with $0 < k' < k$,

$$\lim_{n\to\infty} \int_M \frac{\varphi_n \mu}{1 - |\mu|^2}\, \mathrm{d}x\, \mathrm{d}y = \frac{k}{1 - k^2}.$$

□

We are now in a position to prove the sufficiency of the Hamilton–Krushkal condition.

Theorem 8.4.3. *If $\mu \in B(M)$ is a Beltrami differential which is infinitesimally extremal, then μ is extremal in its Teichmüller class.*

Proof. Recall that if μ is infinitesimally extremal, then $||\Lambda_\mu|| = ||\mu||_\infty$. Lemma 8.4.2 implies that $I(\mu) = k(1-k^2)^{-1}$, and then Theorem 8.4.2 gives the result. □

Example 8.4.1. If Ω is a plane domain, and $a \in \mathbb{C}$ is a complex constant with $0 < |a| < 1$, then

$$f(z) = z + a\overline{z}$$

for $z \in \Omega$ is an affine stretch of Ω by the factor

$$K_f = (1+|a|)(1-|a|)^{-1}$$

in the direction of $1/2 \arg a$. Since $\mu_f \equiv a$, the necessary and sufficient condition that f is extremal for its boundary values is that

$$||\Lambda_1|| = \sup \left| \int_\Omega \varphi(z)\, \mathrm{d}x\, \mathrm{d}y \right| = 1, \tag{8.20}$$

where the supremum is taken over all $\varphi \in A^1(\Omega)$ with $||\varphi||_1 = 1$. When $|\Omega| < \infty$, condition (8.20) is satisfied, but it turns out that even when $|\Omega| = \infty$, f being extremal is independent of the value of a and, in particular, of the direction of the stretch.

Consider the case of the affine stretch of the strip $S = \{x + iy : -1 < y < 1\}$, which we know from Example 8.1.1 to be extremal. To prove this via condition (8.20), let

$$\sigma_n(z) = \frac{1}{n} \mathrm{e}^{-z^2/n^2}.$$

We then have that

$$\int_S \sigma_n(z)\, \mathrm{d}x\, \mathrm{d}y = 2\sqrt{\pi},$$

and also

$$\int_S |\sigma_n(z)|\, \mathrm{d}x\, \mathrm{d}y = \sqrt{\pi} \int_{-1}^{1} \mathrm{e}^{y^2/n^2}\, \mathrm{d}y \to 2\sqrt{\pi}.$$

Therefore, $||\Lambda_1|| = 1$, and condition (8.20) is satisfied.

We can use the Hamilton–Krushkal condition to prove the famous Teichmüller's existence theorem.

Theorem 8.4.4. *Let M be a Riemann surface of finite analytic type. Then every Teichmüller equivalence class has a uniquely extremal representative, which is of Teichmüller type.*

Proof. Let $[\mu] \in T(M)$, suppose that μ is extremal in $[\mu]$, and consider the linear functional on $A^1(M)$,

$$\Lambda_\mu(\varphi) = \int_M \mu\varphi$$

for all $\varphi \in A^1(M)$. By Theorem 7.7.7, the dimension of $A^1(M)$ is finite, and so there exists $\varphi_0 \in A^1(M)$ such that Λ_μ achieves its norm at φ_0. By Theorem 8.2.1, μ is also infinitesimally extremal, and therefore

$$||\Lambda_\mu|| = \int_M \mu\varphi_0 = ||\mu||_\infty. \tag{8.21}$$

By (8.21), the only possibility for μ is that $\mu = \overline{\varphi_0}/|\varphi_0|$, and by Theorem 8.3.3, this means that μ is uniquely extremal. □

Exercise 8.4.1. Let M be a Riemann surface such that dim $A^1(M) = \infty$. Prove that there exists $[\mu] \in T(M)$ such that $[\mu]$ does not contain a representative of Teichmüller type. (Hint: use the Hamilton–Krushkal condition to transfer the problem to one of linear functionals, and show that if dim $A^1(M) = \infty$, then there exists a linear functional on $A^1(M)$ that does not attain its norm.)

Exercise 8.4.2. A Riemann surface M is said to be of bounded geometry if there are no arbitrarily short geodesics in M. Show that $A^1(M) \subseteq Q(M)$ if M is of bounded geometry. (Hint: show that if every linear functional on $A^1(M)$ attains its norm, then $A^1(M)$ is reflexive.) The converse of this problem is proved in [39].

9

Unique extremality

We investigate conditions for unique extremality, including the frame mapping theorem, which gives a condition on when a Teichmüller equivalence class has a uniquely extremal representative of Teichmüller type. We will see that unique extremality corresponds to unique infinitesimal extremality. Uniquely extremal mappings need not be of Teichmüller type, and indeed that they need not have constant modulus.

9.1 The frame mapping condition

A Beltrami coefficient μ is uniquely extremal in its Teichmüller class if for all $\nu \sim \mu$, we have that $||\mu||_\infty < ||\nu||_\infty$. The main tool that we will use to find conditions for unique extremality is the Main Inequality, recalling Theorem 8.3.3, where we proved that a Beltrami differential of Teichmüller type is uniquely extremal.

In this section we prove frame mapping condition, which tells us when a Teichmüller class has an extremal element of Teichmüller type, and hence uniquely extremal. In order to state the condition, we need to define the boundary dilatation $H(f)$ of a quasiconformal mapping f of a Riemann surface M. For every compact set $E \subset M$, let $K(M \setminus E, f)$ be the maximal dilatation of f on the complement of E, and let $K_0(M \setminus E, f)$ be the infimum of $K(M \setminus E, f_1)$, where f_1 is any quasiconformal mapping in the same Teichmüller class as f. The boundary dilatation $H(f)$ is the limit of the numbers $K_0(M \setminus E, f)$ as E increases to M. If M is compact, or compact with a finite number of punctures, then $H(f) = 1$ for every quasiconformal mapping f of M. Thus, the boundary dilatation is only interesting when the Teichmüller space of M is infinite dimensional. We can now state the frame mapping condition (see [50]).

Theorem 9.1.1. *Let $f : M \to f(M)$ be a quasiconformal mapping of a Riemann surface M. Let $K_0(f)$ denote the infimum of the maximal dilatations $K(f_1)$ over all quasiconformal mappings f_1 which are in the Teichmüller* *equivalence class of f. Suppose that $H(f) < K_0(f)$. Then the Teichmüller equivalence class of f is represented by a quasiconformal mapping f_0, whose Beltrami coefficient is of Teichmüller type,*

$$k\frac{\bar{\varphi}}{|\varphi|},$$

where $0 < k < 1$, and φ is an integrable, holomorphic, quadratic differential on M. Further, k and φ (up to multiplication by a positive constant) are uniquely determined by the Teichmüller class of f, and $k\bar{\varphi}/|\varphi|$ is the Beltrami coefficient of the unique extremal representative of the Teichmüller class of f.

Proof. Let f_0 and f_1 be equivalent quasiconformal mappings on M, with respective Beltrami coefficients μ_0 and μ_1. Assuming $||\varphi||_1 = 1$, we can write the Main Inequality (8.11) as

$$1 \leq \int_M |\varphi(z)| \frac{|1-\mu_0(z)\varphi(z)/|\varphi(z)||^2}{1-|\mu_0(z)|^2} \cdot \frac{|1+(\nu_1/\nu_0)\mu_0\alpha\varphi(z)/|\varphi(z)||^2}{1-|\mu_1(z)|^2}\, dx\, dy, \tag{9.1}$$

where

$$\alpha = \frac{1-\overline{\mu_0\varphi}/|\varphi|}{1-\mu_0\varphi/|\varphi|}$$

and ν_1 and ν_0 are, respectively, the Beltrami coefficients of the mappings $(f_1)^{-1}$ and $(f_0)^{-1}$. Note that $||\nu_1||_\infty = ||\mu_1||_\infty$, and that μ_0/ν_0 and α are measurable functions whose absolute value is equal to 1 almost everywhere.

Now assume that f_0 is an extremal quasiconformal mapping in the class of f with Beltrami coefficient μ_0 such that

$$K_0(f) = (1 + ||\mu_0||_\infty)(1 - ||\mu_0||_\infty)^{-1}.$$

Since μ_0 is extremal in its Teichmüller class, it is also infinitesimally extremal. Then a Hamilton sequence φ_n for μ_0 satisfies

$$\int_M |\varphi_n| - \Re \int_M \frac{\mu_0}{||\mu_0||_\infty}\varphi_n \to 0,$$

as $n \to \infty$, and $||\varphi_n||_1 = 1$ for all n. First we need to show that the hypothesis that $H(f) < K_0(f)$ implies that μ_0 cannot have a degenerating Hamilton sequence. A Hamilton sequence is called degenerating if it converges uniformly to 0 on compact subsets of M. If M is compact, there cannot be any degenerating Hamilton sequences, because the locally uniform convergence of the normal family φ_n (in local coordinates) would then imply uniform convergence. Therefore the claim that μ_0 has no degenerating Hamilton sequences is only of interest in the case where M is not compact.

If $H(f_0) < K_0(f_0) = K_0$, then there exists a quasiconformal mapping f_1 equivalent to f_0 which has dilatation H_1 on the complement subset E of M, where $H_1 < K_0$. The mapping f_1 is called a frame mapping for f_0. The maximal dilatation of f_1 on the whole of M is bounded by some constant K_1, which may be much larger than K_0. The inequality (9.1) then becomes

$$1 \leq K_1 K_0 \int_E |\varphi_n| \, dx \, dy + H_1 \int_{M \setminus E} \frac{|1 - \mu_0 \varphi_n / |\varphi_n||^2}{1 - |\mu_0|^2} \, dx \, dy. \tag{9.2}$$

If we assume that φ_n is a degenerating Hamilton sequence, then the first term on the right-hand side of (9.2) converges to 0 as $n \to \infty$. Let $k_0 = ||\mu_0||_\infty$ and $K_0 = (1 + k_0)(1 - k_0)^{-1}$. Using Lemmas 8.4.1 and 8.4.2, we obtain

$$\lim_{n \to \infty} \int_M |\varphi_n| \frac{|\mu_0|}{1 - |\mu_0|^2} = \frac{k_0^2}{1 - k_0^2}$$

and

$$\lim_{n \to \infty} \int_M |\varphi_n| \frac{\varphi_n \mu_0}{1 - |\mu_0|^2} = \frac{k_0}{1 - k_0^2}.$$

Using these two equations, and the definition of φ_n being a Hamilton sequence, we see that the second integral on the right-hand side of (9.2) converges to K_0^{-1}. Thus, we get $1 \leq H_1 K_0^{-1}$ which contradicts the hypothesis that $H_1 < K_0$. Therefore, μ_0 cannot have a degenerating Hamilton sequence.

We next have to show that a Hamilton sequence φ_n for μ_0 must converge in the L^1 norm to a quadratic differential φ with $||\varphi||_1 = 1$. The sequence φ_n must have a convergent subsequence, since it is a normal family of holomorphic functions, in local coordinates. We now assume φ_n to be the subsequence. The limit quadratic differential φ must have $||\varphi||_1 \leq 1$. Since μ_0 cannot have any degenerating Hamilton sequences, then $||\varphi||_1 \neq 0$. If $||\varphi||_1 < 1$, then we could construct another Hamilton sequence

$$\psi_n = \frac{\varphi_n - \varphi}{||\varphi_n - \varphi||_1}$$

which would be degenerating. Since μ_0 cannot have a degenerating Hamilton sequence, we conclude that we must have $||\mu\varphi||_1 = 1$. Now, we have the inequality

$$\int_M |\varphi_n - \varphi| \leq ||\varphi_n||_1 - ||\varphi||_1 + 2 \int_{M \setminus E} |\varphi| + 2 \int_E |\varphi_n - \varphi|.$$

Since the norms of φ_n and φ are equal to 1, this implies that φ_n converges to φ in the L^1 norm. From the necessity of the Hamilton–Krushkal condition, we know that a Hamilton sequence must exist, and so must converge in norm to φ with $||\varphi||_1 = 1$. Therefore,

$$\int_M \varphi \mu_0 \, dx \, dy = ||\mu_0||_\infty = k_0,$$

which is only possible if $\mu_0 = k_0 \bar{\varphi} / |\varphi|$. By Theorem 8.3.3, μ_0 is unique, which proves the theorem. □

The frame mapping theorem applied to Riemann surfaces of finite analytic type shows that there is always a unique extremal representative in every Teichmüller class, which gives another proof of Theorem 8.4.4.

Definition 9.1.1. *An element* $[\mu] \in T(M)$ *is called a Strebel point if the Teichmüller class of* μ *contains a frame mapping.*

The following material on Strebel points, and more, can also be found in [22].

The frame mapping theorem has an analogue for infinitesimal equivalence classes. We first need to define an appropriate analogue to a frame mapping. For a Beltrami differential μ on M, let $b^*(\mu)$ be defined by

$$\inf \sup_{z \in M \setminus K} \{|\mu(z)|\},$$

where the infimum is taken over all compact subsets K of M. The infinitesimal boundary dilatation $b(\Lambda)$ of a linear functional Λ on $A^1(M)$ is the infimum of $b^*(\mu)$ taken over the infinitesimal equivalence class of μ.

Definition 9.1.2. *A linear functional* Λ *on* $A^1(M)$ *is called a Reich–Strebel functional if* $b(\Lambda) < ||\Lambda||$.

We can now prove the infinitesimal version of the frame mapping theorem.

Theorem 9.1.2. *Let* Λ *be a Reich–Strebel functional. Then there exists a unique* $\varphi \in A^1(M)$ *with norm one such that* $\Lambda = \Lambda_\mu$ *where* $\mu = ||\Lambda||\bar{\varphi}/|\varphi|$.

Proof. We know that from the Hahn–Banach theorem and the Riesz representation theorem that $\Lambda = \Lambda_\mu$ where $||\mu||_\infty = ||\Lambda||$. Now, there exists a sequence φ_n in the unit ball of $A^1(M)$ such that

$$||\Lambda|| = \lim_{n \to \infty} \Lambda(\varphi_n). \tag{9.3}$$

Supposing there exists $\varphi \in A^1(M)$ of norm one such that $\Lambda(\varphi) = ||\Lambda||$, then

$$0 = ||\Lambda|| ||\varphi|| - \Lambda(\varphi) = \int_M (||\Lambda|| ||\varphi|| - \mu\varphi).$$

The only possibility is to have $\mu = ||\Lambda||\bar{\varphi}/|\varphi|$. If ψ also satisfies $\Lambda(\psi) = ||\Lambda||$, then this reasoning shows that $\mu = ||\Lambda||\bar{\psi}/|\psi|$. Since φ and ψ both have norm one, then $\varphi = \psi$ and the extremal μ such that $\Lambda = \Lambda_\mu$ is uniquely determined.

So now we have to show that if $b(\Lambda) < ||\Lambda||$, then every such sequence in (9.3) for Λ converges to $\varphi \in A^1(M)$ such that $\Lambda(\varphi) = ||\Lambda||$. Since φ is uniquely determined, it is enough to show that any sequence in (9.3) has a convergent subsequence. We can also assume that φ_n converges locally uniformly to φ and that $||\varphi_n|| = 1$ for all n.

Since $0 \leq |\varphi_n - \varphi| + |\varphi| - |\varphi_n| \leq 2|\varphi|$, Lebesgue's dominated convergence theorem implies

$$\lim_{n \to \infty} (||\varphi_n - \varphi|| + ||\varphi|| - ||\varphi_n||) = \int_M \lim_{n \to \infty} (|\varphi_n - \varphi| + |\varphi| - |\varphi_n|) = 0.$$

Thus

$$\lim_{n \to \infty} ||\varphi_n - \varphi|| = 1 - ||\varphi||.$$

It is clear that $||\varphi|| \leq 1$. In the case where $||\varphi|| = 0$, then this implies that φ_n is a degenerating sequence and

$$b(\Lambda) \geq \limsup_{n\to\infty} |\Lambda(\varphi_n)| = \Lambda(\varphi),$$

which contradicts Λ being a Reich–Strebel functional. If $0 < ||\varphi|| < 1$, then let

$$\psi_n = \frac{\varphi_n - \varphi}{||\varphi_n - \varphi||}.$$

Then ψ_n is degenerating, and

$$b(\Lambda) \geq \limsup_{n\to\infty} \frac{|\Lambda(\psi_n)|}{||\psi_n||} \geq \frac{||\Lambda|| - |\Lambda(\varphi)|}{1 - ||\varphi||} \geq \frac{||\Lambda|| - ||\Lambda|| ||\varphi||}{1 - ||\varphi||} = ||\Lambda||,$$

again contradicting Λ being a Reich–Strebel functional. Finally, $||\varphi|| = 1$ and so $||\varphi_n - \varphi|| \to 0$ and $\Lambda(\varphi) = ||\Lambda||$, which completes the proof. □

The following lemma will be needed to derive some useful inequalities.

Lemma 9.1.1. *Let the quasiconformal map f with Beltrami differential μ be extremal in its Teichmüller class. Let $0 < \epsilon < 1$ be given and E_n an increasing sequence of compact subsets of M such that $\bigcup_{n=1}^{\infty} E_n = M$. Let*

$$\mu_n(z) = \begin{cases} \mu(z), & z \in E_n, \\ (1-\epsilon)\mu(z), & \textit{otherwise}, \end{cases}$$

and let f_n be quasiconformal with Beltrami differential μ_n. Then, for large enough n, f_n is a frame mapping for its Teichmüller class and the dilatation of an extremal map in the class of f_n approaches the dilatation of f.

Proof. Clearly f_n converges locally uniformly to f. Let g_n be an extremal map in the class of f_n, and let K_n be the dilatation of g_n. If $K_n \leq K_0 - \delta$ for some $\delta > 0$ and for all n, where K_0 is the dilatation of f, then any limit g which a subsequence of the g_n converges to locally uniformly must have dilatation at most $K_0 - \delta$. However, g must be in the Teichmüller class of f, which is itself extremal and thus is a contradiction. Since f_n has dilatation at most

$$\frac{1 + k_0 - \epsilon}{1 - k_0 + \epsilon}$$

on $M \setminus E_n$, for large enough n, f_n is a frame mapping in its class. □

Theorem 9.1.3. *Let $f : M \to N$ be quasiconformal, and K_0 be the dilatation of an extremal map in the Teichmüller class of f. Then*

$$\frac{1}{K_0} \leq \int_M \frac{|1 - \mu\bar{\varphi}/|\varphi||^2}{1 - |\mu|^2} |\varphi| \, dx \, dy, \tag{9.4}$$

for all $\varphi \in A^1(M)$ of norm one, and

$$K_0 \leq \sup \int_M \frac{|1+\mu\bar{\varphi}/|\varphi||^2}{1-|\mu|^2}|\varphi|\, dx\, dy, \tag{9.5}$$

where the supremum is over all $\varphi \in A^1(M)$ of norm one.

Proof. Let f_0 be an extremal representative of its class with Beltrami differential μ_0. Let $\widetilde{\mu}(f_0)$ be the Beltrami differential of f_0^{-1}. Since $\widetilde{\mu} \circ f_0 = -\mu_0\bar{p}/p$, where $p = (f_0)_z$, we have $||\widetilde{\mu}||_\infty = ||\mu_0||_\infty = k_0$. Thus the second expression of the Main Inequality (8.11) satisfies

$$\frac{|1+\widetilde{\mu}(f_0)\theta\bar{\varphi}/|\varphi||^2}{1-|\widetilde{\mu}(f_0)|^2} \leq \frac{1+k_0}{1-k_0}.$$

Therefore we get (9.4) for every $\varphi \in A^1(M)$ of norm one.

For (9.5), consider Lemma 9.1.1. By the frame mapping theorem applied to the Teichmüller class of f_n, there exists $\varphi_n \in A^1(M)$ of norm one such that

$$K_n \leq \int_M \frac{|1+\mu_n\overline{\varphi_n}/|\varphi_n||^2}{1-|\mu_n|^2}|\varphi_n|\, dx\, dy.$$

Now, for any $z, w \in \mathbb{C}$ with $|z|, |w| \leq k_1 < 1$, we have

$$\left|\frac{|1+z|^2}{1-|z|^2} - \frac{|1+w|^2}{1-|w|^2}\right| \leq \frac{7|z-w|}{(1-k_1)^2}.$$

Since $||\mu_n - \mu||_\infty \leq \epsilon$,

$$K_n - \epsilon' \leq \int_M \frac{|1+\mu\overline{\varphi_n}/|\varphi_n||^2}{1-|\mu|^2}|\varphi_n|\, dx\, dy,$$

where $\epsilon' = 7\epsilon(1-k_1)^{-2}$. Letting $n \to \infty$ gives

$$K_0 - \epsilon' \leq \limsup_{n\to\infty} \int_M \frac{|1+\mu\overline{\varphi_n}/|\varphi_n||^2}{1-|\mu|^2}|\varphi_n|\, dx\, dy$$

Since ϵ' is arbitrary, this completes the proof. □

The material on Strebel points is due to Lakic in [30].

Theorem 9.1.4. *Let μ be an extremal Beltrami differential. Then the following are equivalent:*

(a) *μ represents a Strebel point,*

(b) *Λ_μ is a Reich–Strebel functional,*

(c) *μ has no degenerating Hamilton sequence.*

Proof. We first show that (b) is equivalent to (c). The proof of the infinitesimal version of the frame mapping theorem, Theorem 9.1.2, shows that if $b(\Lambda_\mu) < ||\Lambda_\mu||$, then Λ_μ does not have a degenerating Hamilton sequence. For the converse, assume that Λ_μ is not a Reich–Strebel functional, and

$$b(\Lambda_\mu) = ||\Lambda_\mu|| = ||\mu||_\infty.$$

Let E_n be an increasing sequence of compact subsets of M such that

$$\bigcup_{n=1}^{\infty} E_n = M.$$

For each n, let

$$\mu_n(z) = \begin{cases} \mu(z), & z \notin E_n, \\ 0, & z \in E_n. \end{cases}$$

Now $b^*(\mu - \mu_n) = 0$, and so $b([\mu]) = b([\mu_n])$. Therefore,

$$||\Lambda_\mu|| = b([\mu]) = b([\mu_n]) \le ||[\mu_n]|| \le ||\mu_n||_\infty \le ||\mu||_\infty = ||\Lambda_\mu||.$$

This implies that μ_n is infinitesimally extremal, and so there exists $\varphi_n \in A^1(M)$ of norm one such that

$$\int_M \mu_n \varphi_n \ge ||\mu||_\infty - \frac{1}{n}.$$

Therefore

$$\int_{M \setminus E_n} |\varphi_n| \to 1$$

as $n \to \infty$. This means that φ_n is degenerating and

$$\int_M \mu \varphi_n \to ||\mu||_\infty.$$

This shows that (b) and (c) are equivalent. The proof of the frame mapping theorem, Theorem 9.1.1, shows that if μ represents a Strebel point, then μ does not have any degenerating Hamilton sequences. Therefore we need to show that if the Teichmüller class of μ contains no frame mappings, then μ has a degenerating sequence. If μ is identically zero, then any degenerating sequence is a Hamilton sequence for μ. We can therefore assume that $||\mu||_\infty > 0$.

Since the Teichmüller class of μ does not contain any frame mappings, this implies that $H(\mu) = K_0(\mu)$, recalling the notation of the frame mapping theorem. Let μ_n and E_n be defined as above. Then

$$H(\mu) = H(f^\mu) = H(f^\mu \circ (f^{\mu_n})^{-1} \circ f^{\mu_n}) \le H(f^{\mu_n}) = H(\mu_n)$$

since $H(f \circ g) \le H(f)H(g)$ and $H(f^\mu \circ (f^{\mu_n})^{-1}) = 1$. Therefore

$$K_0(\mu) = H(\mu) \le H(\mu_n) \le K_0(\mu_n) \le \frac{1 + ||\mu_n||_\infty}{1 - ||\mu_n||_\infty} \le \frac{1 + ||\mu||_\infty}{1 - ||\mu||_\infty} = K_0(\mu).$$

This string of inequalities must therefore all be equalities, which implies that μ_n is extremal in its class, and by (9.5), there exists $\varphi_n \in A^1(M)$ of norm one such that

$$K_0(\mu) - \frac{1}{n} \le \int_M |\varphi_n| \frac{|1+\mu_n\overline{\varphi_n}/|\varphi_n||^2}{1-|\mu_n|^2}\, dx\, dy$$

$$= \int_{E_n} |\varphi_n|\, dx\, dy + \int_{M\setminus E_n} |\varphi_n| \frac{|1+\mu\overline{\varphi_n}/|\varphi_n||^2}{1-|\mu|^2}\, dx\, dy.$$

Therefore

$$K_0(\mu) - \frac{1}{n} \le \int_{E_n} |\varphi_n| + K_0(\mu) \int_{M\setminus E_n} |\varphi_n|.$$

Since $K_0(\mu) > 1$, the only possibility is that φ_n is degenerating. Also, since

$$\int_{E_n} |\varphi_n| \to 0,$$

we get

$$K_0(\mu) \le \limsup_{n\to\infty} \int_M |\varphi_n| \frac{|1+\mu\overline{\varphi_n}/|\varphi_n||^2}{1-|\mu|^2}\, dx\, dy.$$

Manipulating this inequality, we find that

$$k_0 \le \limsup_{n\to\infty} k_0 \int_M \Re(\mu_n\varphi_n)$$

and therefore

$$||\mu||_\infty \le \limsup_{n\to\infty} \int_M \mu\varphi_n,$$

which means that φ_n is a degenerating Hamilton sequence for μ, and completes the proof. □

Theorem 9.1.5. *The set of Strebel points is open and dense in $T(M)$.*

Proof. An element $[\mu]$ of $T(M)$ is a Strebel point if $H([\mu]) < K_0([\mu])$. Since both $H([\mu])$ and $K_0([\mu])$ are continuous on $T(M)$, the set of Strebel points is open. The fact that the set of Strebel points is dense in $T(M)$ follows immediately from Lemma 9.1.1, by taking a large enough compact subset E_n of M and considering the corresponding μ_n. □

9.2 Some necessary conditions for unique extremality

Theorem 9.2.1. *Let $f : \mathbb{D} \to \mathbb{D}$ be uniquely extremal for its boundary values. Then for any open subset U of $\mathbb{D}$,*

$$\sup_{z\in U} D_f(z) = K_f.$$

Proof. The proof runs by contradiction. Assuming that $K_f > 1$, suppose there exists $\epsilon > 0$ and a disk $\{|z - z_0| < \rho\} \subset \mathbb{D}$ such that

$$\sup_{|z-z_0|<\rho} D_f(z) < K_f - \epsilon,$$

where the supremum is the essential supremum, which neglects sets of two dimensional measure zero. We construct a variation of f to contradict the unique extremality of f.

For $|\delta| < 1$ and $|z| < 1$, consider the homeomorphism

$$T(\delta, z) = z + (1 - |z|)\delta$$

of $|z| < 1$ onto itself, which maps 0 to δ, and is the identity on $|z| = 1$. A calculation shows that

$$\mu_T(z) = \frac{\delta z}{\delta \bar{z} - 2|z|}$$

and

$$k_T = \frac{|\delta|}{2 - |\delta|}.$$

Let

$$g(w) = \begin{cases} z_0 + \rho T(\delta, (w - z_0)\rho^{-1}), & |w - z_0| < \rho \\ w, & |w - z_0| \geq \rho, |w| < 1, \end{cases}$$

and define $\widetilde{f} = f \circ g$. Then $\widetilde{f}$ has the same boundary values as f, and if $|\delta|$is small enough, then $K_f = K_{\widetilde{f}}$, which contradicts the fact that f is uniquely extremal. □

This theorem shows that, in some sense, if f is uniquely extremal, then D_f must equal K_f in most places. If f is uniquely extremal, then the restriction of f to any subregion is also uniquely extremal, and, in particular, for its boundary values on that subregion. However, the converse is not true. Even if the restriction of a mapping to every subregion is extremal for the boundary values on that subregion, the original mapping need not be uniquely extremal on the original region.

Theorem 9.2.2. *Suppose that f is a uniquely extremal quasiconformal mapping of $\mathbb{D}$, and*

$$\mu_f(z) = |\mu_f(z)| \frac{\overline{\varphi(z)}}{|\varphi(z)|}$$

where $\varphi(z)$ equals a holomorphic function $\psi(z)$ in $\mathbb{D} \setminus K$, for a compact subset K of $\mathbb{D}$. Then ψ has a holomorpic extension to all of $\mathbb{D}$, and

$$\mu_f(z) = ||\mu_f||_\infty \frac{\overline{\psi(z)}}{|\psi(z)|}.$$

Proof. Let (A_n),

$$A_n = \{z \in \mathbb{D} : \rho_n \leq |z| \leq \rho'_n\}$$

be a sequence of closed annuli in $\mathbb{D}$, with $\rho_n < \rho'_n < \rho_{n+1}$ with $\rho_n \to 1$ as $n \to \infty$, such that $\psi(z) \neq 0$ in A_n for $n = 1, 2, \ldots$. By the remark preceding this theorem, the restriction of f to $\{z : |z| < \rho_n\}$ is extremal for its boundary values on $\{|z| = \rho_n\}$. Applying Example 8.2.2 to the restriction of f for successive n then gives the theorem. □

9.3 Delta inequalities

The remainder of this chapter is based on material from [8]. In this section, we use the Main Inequality to derive the Delta Inequality and its infinitesimal version. Let f and g be quasiconformal mappings on M in the same Teichmüller class with Beltrami differentials μ and ν respectively. Let $\widetilde{\mu}$ and $\widetilde{\nu}$ be the Beltrami differentials of f^{-1} and g^{-1}.

Theorem 9.3.1. *If μ and ν are in the same Teichmüller equivalence class with*

$$||\nu||_\infty \leq k = ||\mu||_\infty < 1,$$

then

$$\int_M \left| \frac{\widetilde{\mu}(f) - \widetilde{\nu}(f)}{1 - \overline{\widetilde{\mu}(f)}\widetilde{\nu}(f)} \right|^2 |\varphi| \leq C \left(k||\varphi|| - \Re \int_M \mu\varphi \right)$$

for all $\varphi \in A^1(M)$. The constant C depends only on k.

Proof. Let $\alpha(z) = \widetilde{\mu}(f(z))$ and $\beta(z) = \widetilde{\nu}(f(z))$, $\rho = |\alpha - \beta|^2$,

$$I = I(\varphi) = \int_M \frac{\rho}{|1 - \bar{\alpha}\beta|^2} |\varphi|,$$

and $\delta = \delta(\varphi) = k||\varphi|| - \Re \int_M \mu\varphi$. Also let

$$Q = (1 - |\alpha|^2)(1 - |\beta|^2)$$

and

$$P = (\alpha - \beta)(1 - \alpha\bar{\beta})\alpha^{-1}.$$

Suppose first that $|\mu(z)|$ is bounded from below by a positive constant s for almost every $z \in M$. Then for any $\varphi \in A^1(M)$, the Main Inequality implies that

$$\Re \int_M \frac{P}{Q} \mu\varphi \leq \int_M \frac{\rho}{Q} |\varphi|.$$

To get an estimate involving $\delta(\varphi)$, add

$$\int_M |\alpha| \Re \frac{P}{Q} |\varphi|$$

to both sides to get

$$\int_M \frac{|\alpha|\Re P - \rho}{Q}|\varphi| \leq \Re \int_M \frac{P}{Q}(|\alpha||\varphi| - \mu\varphi). \tag{9.6}$$

Furthermore,

$$\Re P = |\alpha|^{-2}\Re((\alpha - \beta)(\bar{\alpha} - |\alpha|^2\bar{\beta})) = |\alpha|^{-2}(\rho + (1 - |\alpha|^2)(\Re\alpha\bar{\beta} - |\beta|^2)).$$

Thus

$$\Re P - \rho = |\alpha|^{-2}(1 - |\alpha|^2)(|\alpha|^2 - \Re\alpha\bar{\beta}),$$

and

$$|\alpha|\Re P - \rho = A\rho + B(|\alpha| - |\beta|) \tag{9.7}$$

where

$$A = (2|\alpha|)^{-1}(1 - |\alpha|)^2$$

and

$$B = (2|\alpha|)^{-1}(|\alpha| + |\beta|)(1 - |\alpha|^2).$$

Since $|\mu|$ is bounded from below by a positive constant $s > 0$, A and B are bounded away from 0 and infinity. Using (9.7) and the estimate $|\beta| - |\alpha| \leq k - |\mu|$, (9.6) yields a positive constant c which depends only on $k = ||\mu||_\infty$ and s such that

$$I \leq c(I_1 + \tau(\varphi))$$

where

$$I_1 = \int_M \sqrt{\rho}||\mu\varphi| - \mu\varphi|$$

and

$$\tau = \tau(\varphi) = \int_M (k - |\mu|)|\varphi|.$$

Using the Cauchy–Schwarz inequality and the identity

$$||w| - w|^2 = 2|w|(|w| - \Re w),$$

we get

$$I_1 \leq c_1 I^{1/2}\delta^{1/2}.$$

Using this and the inequality $\tau \leq \delta$, we see that

$$I \leq c_2(I^{1/2}\delta^{1/2} + \delta).$$

On letting x be the positive number $I^{1/2}\delta^{1/2}$, this inequality becomes

$$x \leq c_2\left(1 + \frac{1}{x}\right),$$

which implies that x is bounded. We conclude that

$$I \leq C(k,s)\delta.$$

Now assume that μ and ν are two arbitrary equivalent Beltrami differentials such that

$$||\nu||_\infty \leq ||\mu||_\infty = k < 1.$$

Note that $I(\varphi)$ is invariant under post-compositions by quasiconformal mappings. Fix $q \in A^1(M)$ of norm one and let

$$E = \left\{ z \in f(M) : |\widetilde{\mu}(z)| < \frac{k}{2} \right\}.$$

Take $z \in E$. If $\widetilde{\mu}(z) \neq \widetilde{\nu}(z)$, let $\eta(z)$ be the point on the hyperbolic circle

$$S = \left\{ w \in \mathbb{D} : d_h(w, \widetilde{\mu}(z)) = \frac{1}{2} d_h \left(0, \frac{k}{2} \right) \right\}$$

with minimal distance from $\widetilde{\nu}(z)$, where d_h is the hyperbolic metric on $\mathbb{D}$. If $\widetilde{\mu}(z) = \widetilde{\nu}(z) \neq 0$, let $\eta(z) = a\widetilde{\mu}(z)$, where a is a positive constant so that $\eta(z) \in S$. Finally, if $\widetilde{\mu}(z) = \widetilde{\nu}(z) = 0$, let

$$\eta(z) = \frac{k|q(z)|}{2q(z)}.$$

Note that this definition is independent of the local coordinate chart containing the point z. Extend η by setting $\eta(z) = 0$ for all z in $f(M) \setminus E$. Let h be a quasiconformal mapping with domain $f(M)$ and Beltrami differential η. Then the Beltrami differentials μ_1 and ν_1 of $h \circ f$ and $h \circ g$ are equivalent. Furthermore,

$$|\mu_1(z)| = \left| \frac{\eta(f(z)) - \widetilde{\mu}(f(z))}{1 - \overline{\eta(f(z))}\widetilde{\mu}(f(z))} \right|$$

and

$$|\nu_1(z)| = \left| \frac{\eta(g(z)) - \widetilde{\nu}(g(z))}{1 - \overline{\eta(g(z))}\widetilde{\nu}(g(z))} \right|.$$

Hence

$$d_h(0, \mu_1) = d_h(\eta(f), \widetilde{\mu}(f))$$

and

$$d_h(0, \nu_1) = d_h(\eta(g), \widetilde{\nu}(g)).$$

The definition of $\eta(z)$ gives $|\mu_1(z)| \geq k/2$ and $|\nu_1(z)| \leq k$ for almost all $z \in M$. Also $|\mu_1(z)| = k/2$ for almost all $z \in f^{-1}(E)$ and $|\mu_1(z)| = |\mu(z)| \geq k/2$ for all $z \in M \setminus f^{-1}(E)$. Since

$$(h \circ g)^{-1} \circ (h \circ f) = g^{-1} \circ f,$$

we have

$$I(\varphi) \le C(k, k/2)\left(k\|\varphi\| - \Re\int_M \mu_1\varphi\right).$$

Let $G = f^{-1}(E)$. Then

$$3\Re\int_G \mu\varphi - \Re\int_G \mu_1\varphi \le 2k\int_G |\varphi|.$$

Hence

$$3\Re\int_M \mu\varphi \le 2k\int_{M\setminus G} |\varphi| + \Re\int_{M\setminus G} \mu\varphi + 2k\int_G |\varphi| + \Re\int_G \mu_1\varphi,$$

which gives

$$k\|\varphi\| - \Re\int_M \mu_1\varphi \le 3\left(k\|\varphi\| - \Re\int_M \mu\varphi\right) = 3\delta(\varphi).$$

Therefore, $I(\varphi) \le 3C(k, k/2)\delta(\varphi)$, which completes the proof. □

We now give an infinitesimal version of the preceding theorem. The proof is an extended calculation, and as such is omitted, but can be found in [8].

Theorem 9.3.2. *There exists a universal constant C such that for every pair μ and ν of infinitesimally equivalent Beltrami differentials with*

$$\|\nu\|_\infty \le \|\mu\|_\infty < \infty,$$

we have

$$\int_M |\mu - \nu|^2|\varphi| \le C\|\mu\|_\infty\left(\|\mu\|_\infty\|\varphi\| - \Re\int_M \mu\varphi\right),$$

for all $\varphi \in A^1(M)$. The constant C is independent of μ and ν.

9.4 Beltrami differentials with constant modulus

A natural question to ask is whether there is an analogue to the Hamilton–Krushkal condition for unique extremality. That is, whether μ is uniquely extremal in its Teichmüller class if and only if it is uniquely infinitesimally extremal. The answer is in the affirmative, and we begin by showing that this is the case in the situation where μ has constant absolute value.

Lemma 9.4.1. *If μ has constant absolute value and is uniquely extremal in its Teichmüller class, then it is uniquely infinitesimally extremal.*

Proof. Let μ be uniquely extremal in its Teichmüller class with $|\mu(z)| = k$ for all z. Suppose for a contradiction that μ is not uniquely infinitesimally extremal.

Then there exists $\nu \in [\mu]_*$ such that $||\nu||_\infty \leq k$. Recall the Infinitesimal Delta Inequality in Theorem 9.3.2,

$$\int_M |\mu - \nu|^2 |\varphi| \leq C \left(k||\varphi|| - \Re \int_M \mu\varphi \right), \tag{9.8}$$

for all $\varphi \in A^1(M)$. Since μ and ν are distinct, there exists $\epsilon > 0$ and a compact set $E \subseteq M$ of positive Lebesgue measure such that $|\mu - \nu| \geq \epsilon$ on E. The Infinitesimal Delta Inequality then gives

$$\epsilon^2 \int_E |\varphi| \leq C \left(k - \Re \int_M \mu\varphi \right) \tag{9.9}$$

when $||\varphi|| = 1$. To find a contradiction, we need to show that this inequality fails for some $\varphi \in A^1(M)$ of norm one. As in Lemma 9.1.1, for arbitrary $r > 0$ form the truncation α of μ to E by

$$\alpha(z) = \begin{cases} \mu(z), & z \in E, \\ \frac{\mu(z)}{1+r}, & z \in M \setminus E. \end{cases}$$

Then $[\alpha]$ is a Strebel point, and by the frame mapping theorem, there exists

$$s_r = k_0([\alpha]) > s := k(1+r)^{-1}$$

and $\varphi \in A^1(M)$ of norm one such that α and $s_r\bar{\varphi}/|\varphi|$ are Teichmüller equivalent. By (9.5),

$$\frac{1+s}{1-s} < \frac{1+s_r}{1-s_r} \leq \int_M \frac{|1 + \alpha\bar{\varphi}/|\varphi||^2}{1 - |\alpha|^2} |\varphi|.$$

If $\mu_1 = \mu(1+r)^{-1}$, then

$$\frac{1+s}{1-s} < \int_M \frac{|1 + \mu_1\bar{\varphi}/|\varphi||^2}{1 - |\mu_1|^2} |\varphi| + C_1(k) r \int_E |\varphi|.$$

Manipulating this equation leads to

$$k - \Re \int_M \mu\varphi \leq C_2(k) r \int_E |\varphi|. \tag{9.10}$$

Combining (9.9) with this gives

$$\epsilon^2 \int_E |\varphi| \leq C_3(k) r \int_E |\varphi|,$$

which gives a contradiction if r is sufficiently small. □

Lemma 9.4.2. *If μ has constant absolute value and is uniquely infinitesimally extremal, then it is uniquely extremal in its Teichmüller class.*

Proof. Let μ be uniquely infinitesimally extremal with $|\mu(z)| = k$ for all z. Suppose that μ is not uniquely extremal in its Teichmüller class. Then there exists a Beltrami differential ν distinct from μ in the same Teichmüller class and $||\nu||_\infty \leq k$. Let f and g be quasiconformal mappings of M with Beltrami differentials μ and ν respectively. Let also $\widetilde{\mu}$ and $\widetilde{\nu}$ be the Beltrami differentials of f^{-1} and g^{-1} respectively. Recall the Delta Inequality of Theorem 9.3.1,

$$\int_M \left| \frac{\widetilde{\mu}(f) - \widetilde{\nu}(f)}{1 - \overline{\widetilde{\mu}(f)}\widetilde{\nu}(f)} \right|^2 |\varphi| \leq C \left(k||\varphi|| - \Re \int_M \mu\varphi \right), \tag{9.11}$$

for all $\varphi \in A^1(M)$. Since μ and ν are distinct, there exists $\epsilon > 0$ and a compact set $E \subseteq M$ of positive Lebesgue measure such that $|\widetilde{\mu}(f) - \widetilde{\nu}(f)| \geq \epsilon$ on E. Then, via the Delta Inequality, we have

$$\frac{\epsilon^2}{4} \int_E |\varphi| \leq C \left(k - \Re \int_M \mu\varphi \right)$$

when $||\varphi|| = 1$. The proof now follows the same lines as the previous lemma, defining the truncation α of μ, and this time using the infinitesimal frame mapping theorem. We leave the details to the reader. □

Definition 9.4.1. *For a Beltrami differential μ, the extremal set $X(\mu)$ is defined to be*

$$X(\mu) = \{z \in \mathbb{D} : |\mu(z)| = ||\mu||_\infty\}.$$

Theorem 9.4.1. *Let μ be a Beltrami differential with constant absolute value. Then the following conditions are equivalent:*

(a) *μ is uniquely extremal in its Teichmüller class,*

(b) *μ is uniquely infinitesimally extremal,*

(c) *for every measurable $E \subseteq M$ with non-zero measure, there exists a sequence $\varphi_n \in A^1(M)$, with $||\varphi_n|| = 1$ for all n such that*

$$\frac{1}{\int_E |\varphi_n|} \left(||\mu_\infty|| - \Re \int_M \mu\varphi_n \right) \to 0,$$

(d) *μ is extremal in its Teichmüller class, and for every compact subset $E \subseteq M$ with non-zero measure and every $r > 0$, $[\mu\chi_E + \frac{1}{1+r}\mu\chi_{M\setminus E}]$ is a Strebel point in $T(M)$,*

(e) *μ is uniquely infinitesimally extremal and for every compact subset $E \subseteq M$ with non-zero measure and every $r > 0$,*

$$b\left([\mu\chi_E + \frac{1}{1+r}\mu\chi_{M\setminus E}] \right) < \left|\left| [\mu\chi_E + \frac{1}{1+r}\mu\chi_{M\setminus E}] \right|\right|,$$

(f) *for every admissible variation η of μ, there exists a sequence $\varphi_n \in A^1(M)$ such that*

$$||\varphi_n|| ||\eta||_\infty - \Re \int_M \varphi_n\eta \to 0$$

and

$$\liminf_{n\to 0} |\varphi_n(z)| > 0$$

for almost all $z \in X(\eta)$.

Proof. Lemmas 9.4.1 and 9.4.2 show that (a) and (b) are equivalent. Let (a) hold, let $E \subseteq M$ have positive measure and let $F \subseteq E$ be compact with positive measure. Then inequality (9.10) yields the existence of a sequence $\varphi_n \in A^1(M)$ with $||\varphi_n||_1 = 1$ for all n such that

$$||\mu||_\infty - \Re \int_M \mu\varphi_n \leq \frac{C_2(k)}{n} \int_F |\varphi_n| \leq \frac{C_2(k)}{n} \int_E |\varphi_n|.$$

Furthermore,

$$\frac{1}{\int_E |\varphi_n|}\left(||\mu||_\infty - \Re \int_M \mu\varphi_n\right) \geq \frac{1}{\int_E |\varphi_n|}\left(||\mu||_\infty - ||\mu||_\infty \int_M |\varphi_n|\right) \geq 0$$

for all n, which shows that (a) implies (c). Inequality (9.9) shows that (c) implies (b). The proof of Lemma 9.4.1 subsequent to (9.9) shows that (a) implies (d).

Now suppose that (d) holds. If $[\mu]$ is a Strebel point, then (c) holds for a constant sequence φ_n. If $[\mu]$ is a non-Strebel point, then the proof of Lemma 9.4.1 shows that we have inequality (9.10) and so (d) implies (c). Similarly, Lemma 9.4.2 shows that (b) implies (e) implies (c). Finally, the equivalence of (b) and (d) follows from Lemma 9.6.3 and Exercise 9.6.1 to appear later. □

We now give necessary and sufficient conditions for a Beltrami differential of Teichmüller type to be uniquely extremal.

Theorem 9.4.2. *Let φ be a non-zero quadratic differential on M and f be a quasiconformal mapping defined on M with Beltrami differential $k\bar{\varphi}/|\varphi|$ where $0 < k < 1$. Then f is uniquely extremal if and only if there exists a sequence $\varphi_n \in A^1(M)$ such that*

(a) *$\varphi_n(z)$ converges to $\varphi(z)$ uniformly on compact subsets of M,*

(b) $k||\varphi_n|| - \Re \int_M \varphi_n\mu \to 0$.

Proof. If there exists a sequence satisfying conditions (a) and (b), then by Fatou's lemma, the sequence $\varphi_n/||\varphi_n||$ satisfies condition (c) of Theorem 9.4.1 for every measurable subset E of M. Therefore f is uniquely extremal in its Teichmüller class.

For the converse, suppose f is uniquely extremal and pick a conformal disk E, that is an open set $E \subset M$ such that the closure $\bar{E}$ is in a single coordinate patch, with local coordinate z, and $\bar{E}$ is a closed disk in the z-plane. Since the zeros of φ are discrete, the radius of $z(E)$ can be perturbed so that φ has no zeros on ∂E. By condition (c) of Theorem 9.4.1 there exists a sequence $\psi_n \in A^1(M)$, each with norm one, such that

$$\frac{k - \Re \int_M \psi_n\mu}{\int_E |\psi_n|} \to 0.$$

If

$$\varphi_n = \frac{\psi_n}{\int_E |\psi_n|} \int_E |\varphi|,$$

then $\int_E |\varphi_n| = \int_E |\varphi|$ and

$$0 \le k||\varphi_n|| - \Re \int_M \varphi_n \mu = \int_E \frac{k - \Re \int_M \psi_n \mu}{\int_E |\psi_n|} |\varphi| \to 0.$$

Therefore,

$$0 \le k \int_E |\varphi_n| - \Re \int_E \varphi_n \mu \le k||\varphi_n|| - \Re \int_M \varphi_n \mu \to 0.$$

Since φ is holomorphic in a neighbourhood of $\bar{E}$ and has no zeros on the boundary of E, $A^1(E)$ is weakly uniformly convex at $\varphi / \int_E |\varphi|$. Therefore, $\int_E |\varphi_n - \varphi| \to 0$ and so φ_n converges to φ uniformly on compact subsets of E.

Now let $V_1 = E$ and F be a conformal disk in M such that $F \cap E$ is not empty. Then

$$\begin{aligned} 0 \le k - \Re \int_F \frac{\varphi_n}{\int_F |\varphi_n|} \mu &= \frac{k \int_F |\varphi_n| - \Re \int_F \varphi_n \mu}{\int_F |\varphi_n|} \\ &\le \frac{k \int_M |\varphi_n| - \Re \int_M \varphi_n \mu}{\int_{F \cap E} |\varphi_n|} \to 0 \end{aligned}$$

since $\int_{F\cap E} |\varphi_n| \to \int_{F \cap E} |\varphi| > 0$. So if the boundary of F contains no zeros of φ, then

$$\int_F \left| \frac{\varphi_n}{\int_F |\varphi_n|} - \frac{\varphi}{\int_F |\varphi|} \right| \to 0.$$

Let $p \in F \cap E$, and then

$$\frac{\varphi_n(p)}{\int_F |\varphi_n} - \frac{\varphi(p)}{\int_F |\varphi|} \to 0$$

and $\varphi_n(p) \to \varphi(p)$ in any local coordinate of p. Thus $\int_F |\varphi_n| \to \int_F |\varphi|$. Hence

$$\int_F |\varphi_n - \varphi| \to 0,$$

and φ_n converges uniformly on compact subsets of F. Thus φ_n converges to φ uniformly on compact subsets of $V = V_1 \cup V_2 \ldots$ where V_{n+1} is the union of all conformal disks in M that intersect V_n. Clearly V is open and since every point of M is the centre of some conformal disk in M, V is closed. Therefore V is both open and closed and since M is connected, $V = M$. □

9.5 Beltrami differentials with non-constant modulus

Theorem 9.5.1. *There exists a uniquely extremal mapping f of $\mathbb{D}$ such that $|\mu_f| < k_f$ on a set of positive measure.*

Proof. Choose an arbitrary compact subset K of $\mathbb{D}$ which contains at least two points and such that $\mathbb{D}\setminus K$ is doubly connected, and also choose k such that $0<k<1$. We construct a function $\mu \in L^\infty(\mathbb{D})$ and a sequence $\varphi_n \in A^1(\mathbb{D})$ such that

$$|\mu(z)| = \begin{cases} 0, & z\in K \\ 1, & \text{for almost all } z\in \mathbb{D}\setminus K, \end{cases}$$

$$\lim_{n\to\infty}[||\varphi_n||_1 - \lambda_\mu(\varphi_n)] = 0,$$

and

$$\lim_{n\to\infty}|\varphi_n(z)| = \infty, \text{ almost everywhere in } \mathbb{D}\setminus K.$$

Once this construction is achieved, choose K as a connected subset of $\mathbb{D}$ with empty interior and with $|K|>0$, and set $\mu_f(z) = k\mu(z)$. Then $k_f = k$, and $|\mu_f(z)| < k_f$ for $z\in K$, while by Theorem 9.4.2, f is uniquely extremal.

Now, let $(J_n),(K_n)$ be closed Jordan domains such that

$$J_n\subset\mathbb{D},\quad K_n\subset\mathbb{D},\quad J_n\subset \mathrm{Int}J_{n+1},\quad K_{n+1}\subset\mathrm{Int}K_n,\quad J_n\cap K_n=\emptyset,$$

$$\left|\bigcup_{n=1}^\infty J_n\right| = |\mathbb{D}\setminus K|,\quad \bigcap_{n=1}^\infty K_n = K.$$

The next step is to construct a sequence of polynomials $(Q_n(z))$, a sequence of polynomials $(\varphi_n(z))$ and a subsequence (A_n) of (J_n).

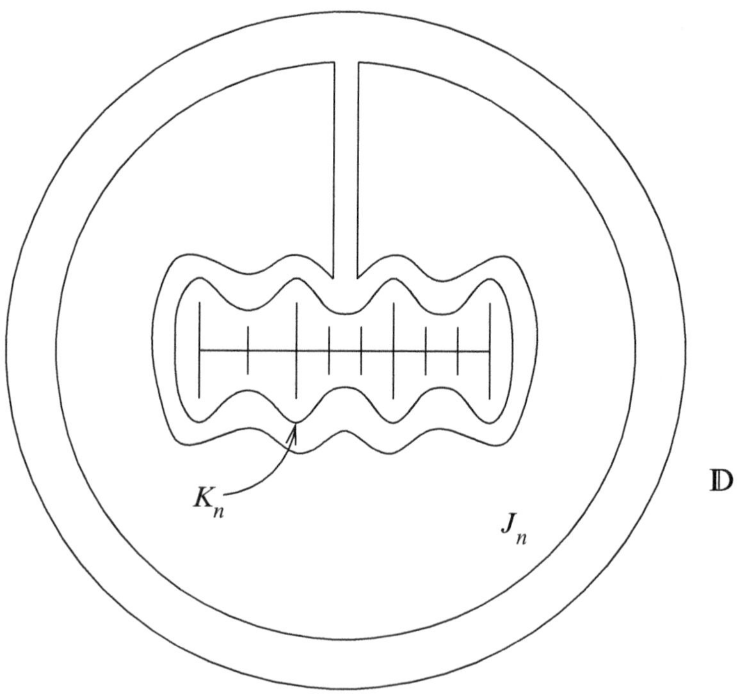

For $n = 1$, let $A_1 = J_1$, and let $Q_1(z)$ be a polynomial formed by Runge's theorem such that $|Q_1(z)| > 2$ when $z \in A_1$, $|Q_1(z)| < 1/2$ when $z \in K$. Set $\varphi_1(z) = Q_1(z)$.

For $n = 2$, choose A_2 from $(J_2, J_3, \ldots)$ such that

$$\int_{\mathbb{D}} |\varphi_1(z)|\, dx\, dy - \int_{K \cup A_2} |\varphi_1(z)|\, dx\, dy < 1.$$

Note that this is possible since $K \cup \bigcup_1^\infty J_n$ has measure π. Let

$$\alpha_1 = \max_{z \in A_2} |\varphi_1(z)|.$$

We use Runge's theorem to find a polynomial $Q_2(z)$ such that when $z \in A_2$, $|Q_2(z)| > 2$ and

$$\alpha_1 \left| 1 - \frac{|Q_2(z)|}{Q_2(z)} \right| < \frac{1}{2},$$

and when $z \in K$,

$$|Q_2(z)| < \frac{1}{2}.$$

Set

$$\varphi_2(z) = \varphi_1(z) Q_2(z).$$

Proceeding inductively, we find Jordan domains $A_3, A_4, \ldots$ from amongst the (J_n), and polynomials $Q_3, Q_4, \ldots$ such that A_{n+1} satisfies

$$\int_{\mathbb{D}} |\varphi_n(z)|\, dx\, dy - \int_{K \cup A_{n+1}} |\varphi_n(z)|\, dx\, dy, \tag{9.12}$$

and such that when $z \in A_{n+1}$, Q_n satisfies $|Q_{n+1}(z)| > 2$ and

$$\alpha_n \left| 1 - \frac{|Q_n(z)|}{Q_n(z)} \right| < \frac{1}{2^n} \tag{9.13}$$

and when $z \in K$, $|Q_{n+1}(z)| < 1/2$, where

$$\alpha_n = \max_{z \in A_{n+1}} |\varphi_n(z)|.$$

Set

$$\varphi_{n+1}(z) = \varphi_n(z) Q_{n+1}(z) \tag{9.14}$$

for all $z \in \mathbb{D}$. For $z \in \cup_1^\infty A_n$, it is clear that

$$\lim_{n \to \infty} |\varphi_n(z)| = \infty$$

and so for almost all $z \in \mathbb{D} \setminus K$, $\lim |\varphi_n(z)| = \infty$ and $\lim ||\varphi_n||_1 = \infty$. Moreover, $|\varphi_n(z)| > 2^n$ for $z \in A_1$. Thus for $n = 1, 2, \ldots$,

$$\alpha_n > 2^n.$$

For $z \in \cup_1^\infty A_n$, let

$$\mu_n(z) = \frac{|\varphi_n(z)|}{\varphi_n(z)}$$

and let $\mu_n(z) = 0$ for $z \in K$. By (9.14),

$$\mu_{n+1}(z) = \mu_n(z)\frac{|Q_{n+1}(z)|}{Q_{n+1}(z)}$$

almost everywhere in $\mathbb{D} \setminus K$. Then by (9.13), for $z \in A_{n+1}$,

$$\alpha_n \left| 1 - \frac{\mu_{n+1}(z)}{\mu_n(z)} \right| > \frac{1}{2^n}.$$

Hence for $z \in A_{n+1}$,

$$\alpha_n|\mu_{n+1}(z) - \mu_n(z)| < 2^{-n}|\mu_n(z)| = 2^{-n},$$

and therefore, for $z \in K \cup A_{n+1}$,

$$|\mu_{n+1}(z) - \mu_n(z)| < 4^{-n}.$$

So for every $z \in K \cup \bigcup_1^\infty A_n$, $(\mu_j(z))$ is a Cauchy sequence, and this implies that $\mu(z) = \lim \mu_n(z)$ exists almost everywhere in $\mathbb{D}$. In particular, we have $|\mu(z)| = 1$ almost everywhere in $\mathbb{D} \setminus K$, and $\mu(z) = 0$ in K. It remains to show that $\lim[||\varphi_n||_1 - \lambda_\mu(\varphi_n)] = 0$. Since the sets A_n are increasing,

$$|\mu(z) - \mu_n(z)| \leq 4^{-n} + 4^{-(n+1)} + 4^{-(n+2)} + \cdots = \frac{1}{3.4^{n-1}},$$

for $z \in K \cup A_{n+1}$. Since (α_n) is an increasing sequence,

$$|\mu - \mu_n| \leq \alpha_n \sum_{j=n}^\infty |\mu_{j+1} - \mu_j| \leq \sum_{j=n}^\infty \alpha_j|\mu_{j+1} - \mu_j| \leq \sum_{j=n}^\infty 2^{-j} = 2^{-(n-1)},$$

for $z \in K \cup A_{n+1}$. That is, for $z \in K \cup A_{n+1}$,

$$|\mu(z) - \mu_n(z)|.|\varphi_n(z)| \leq 2^{-(n-1)}. \tag{9.15}$$

Now by (9.12),

$$||\varphi_n||_1 = \int_{K \cup A_{n+1}} |\varphi_n| \, dx \, dy + o(1),$$

$$\Lambda_\mu(\varphi_n) = \int_{K \cup A_{n+1}} \mu\varphi_n \, dx \, dy + o(1).$$

Now by the definition of μ_n and (9.15),

$$\int_{K \cup A_{n+1}} |\varphi_n| \, dx \, dy = \int_{K \cup A_{n+1}} \mu\varphi_n \, dx \, dy + o(1).$$

It follows that

$$||\varphi_n||_1 = \Lambda_\mu(\varphi_n) + o(1),$$

which completes the construction of $\mu(z)$. □

Note that although $|\mu(z)| = ||\mu||_\infty$ almost everywhere in $\mathbb{D} \setminus K$, Theorem 9.2.2 tells us that μ_f cannot be of Teichmüller type in $\mathbb{D} \setminus K$.

We now sketch how to generalize Lemmas 9.4.1 and 9.4.2 so that the restriction on μ having constant absolute value is lifted.

Lemma 9.5.1. *If μ is uniquely extremal in its Teichmüller class, then it is uniquely infinitesimally extremal.*

Proof. Let μ be uniquely extremal in its Teichmüller class with $||\mu||_\infty = k$ and suppose that μ is not uniquely infinitesimally extremal. Then there exists $\nu \in [\mu]_*$ such that $||\nu||_\infty \leq k$. Hence there exists $\epsilon > 0$ and a compact set $E \subset M$ such that $|\mu(z) - \nu(z)| \geq \epsilon$ for almost all $z \in E$.

Consider the set $F = \{z \in E : |\mu(z)| = k\}$. If F had non-zero measure, then by the proof of Lemma 9.4.1, for every $r > 0$, there exists $\varphi = \varphi_r \in A^1(M)$ of norm one such that

$$k - \Re \int_M \mu\varphi \leq C_2(K) r \int_F |\varphi|. \tag{9.16}$$

Recall the Infinitesimal Delta Inequality of Theorem 9.3.2,

$$\int_M |\mu - \nu|^2 |\varphi| \leq C ||\mu||_\infty \left(||\mu||_\infty ||\varphi|| - \Re \int_M \mu\varphi \right) \tag{9.17}$$

for all $\varphi \in A^1(M)$. This yields

$$\epsilon^2 \int_F |\varphi| \leq Ck \left(k - \Re \int_M \mu\varphi \right)$$

which is contradicts (9.16) for r sufficiently small. Therefore, for some $s < k$, the set $G = \{z \in E : |\mu(z)| \leq s\}$ has positive measure.

If η is any Beltrami differential supported in G with $||\eta||_\infty \leq k$, then there is a positive number r such that the Beltrami differential $\xi(r) = r(\mu\chi_{M \setminus G} + \eta\chi_G)$ is uniquely extremal in its Teichmüller class.

Define η on M in the following way. If $z \notin G$, then $\eta(z) = 0$. If $z \in G$, then let $\eta(z)$ be the point on the circle $\{w : |w - \mu(z)| = k\}$ whose distance to $\nu(z)$ is minimal. Then there exists $r > 0$ such that $r(\mu - \eta)$ is uniquely extremal in its Teichmüller class. Furthermore, $r(\mu - \eta)$ is equivalent to $r(\nu - \eta)$ and

$$||r(\nu - \eta)||_\infty = r||(\nu - \eta)||_\infty \leq rk.$$

Also $|r(\mu(z) - \eta(z))| = rk$ for all $z \in G$. Therefore

$$r(\mu(z) - \eta(z)) = r(\nu(z) - \eta(z))$$

for almost all $z \in G$. Therefore $\mu(z) = \nu(z)$ for almost all $z \in G$, which completes the proof. □

Lemma 9.5.2. *If μ is uniquely infinitesimally extremal, then it is uniquely extremal in its Teichmüller class.*

Proof. Let μ be uniquely infinitesimally extremal with $||\mu||_\infty = k$ and suppose that μ is not uniquely extremal in its Teichmüller class. Then there exists a

Beltrami differential $\nu \in [\mu]$ distinct from μ such that $||\nu||_\infty \leq k$. Let f be a quasiconformal mapping on M with Beltrami coefficient μ. Then there exists a quasiconformal mapping g with domain M and Beltrami differential ν such that f and g are homotopic and agree on the boundary of M. Let $\widetilde{\mu}$ be the Beltrami differential of f^{-1} and $\widetilde{\nu}$ be the Beltrami differential of g^{-1}. Clearly $|\mu| = |\widetilde{\mu}(f)|$ and $|\nu| = |\widetilde{\nu}(g)|$. It suffices to show that $\widetilde{\mu} = \widetilde{\nu}$. If $\widetilde{\mu}$ and $\widetilde{\nu}$ are distinct, then there exists $\epsilon > 0$ and a compact subset $E \subset M$ such that

$$\frac{|\widetilde{\mu}(f(z)) - \widetilde{\nu}(f(z))|}{|1 - \widetilde{\mu}(f(z))\overline{\widetilde{\nu}(f(z))}|} \geq \epsilon$$

for almost all $z \in E$. The proof now follows similar lines to the proof of Lemma 9.5.1. Let $F = \{z \in E : |\mu(z)| = k\}$. If F had non-zero measure, then by the proof of Lemma 9.4.2, there would exist $\varphi = \varphi_r \in A^1(M)$ such that

$$k - \Re \int_M \mu\varphi \leq kr \int_F |\varphi|. \tag{9.18}$$

Recall the Delta Inequality of Theorem 9.3.1,

$$\int_M \left| \frac{|\widetilde{\mu}(f(z)) - \widetilde{\nu}(f(z))|}{|1 - \widetilde{\mu}(f(z))\overline{\widetilde{\nu}(f(z))}|} \right|^2 |\varphi| \leq C\left(k||\varphi|| - \Re \int_M \mu\varphi\right), \tag{9.19}$$

for all $\varphi \in A^1(M)$. Applying this yields

$$\epsilon^2 \int_F |\varphi| \leq Ck\left(k - \Re \int_M \mu\varphi\right),$$

which contradicts (9.18) for sufficiently small r. Therefore, F has measure zero and we can assume that for some $s < k$ the set $G = \{z \in E : |\mu(z)| \leq s\}$ has positive measure.

Now, if η is any Beltrami differential supported on G with $||\eta||_\infty \leq k$, then the Beltrami differential $\xi = \mu\chi_{M\setminus G} + \eta\chi_G$ is uniquely infinitesimally extremal.

Define a Beltrami differential $\eta(z)$ on $f(M)$ in the following way. If $z \notin f(G)$, then $\eta(z) = 0$. If $z \in f(G)$, then let $\eta(z)$ be the point on the hyperbolic circle $\{w \in \mathbb{D} : \rho(w, \widetilde{\mu}(z)) = \rho(0, k)\}$ whose hyperbolic distance to $\widetilde{\nu}(z)$ is minimal. Here we are considering the lift of $f(M)$ to $\mathbb{D}$. Let h be a quasiconformal mapping with domain $f(M)$ and Beltrami differential η. Then the Beltrami differentials μ_1 and ν_1 of $h \circ f$ and $h \circ g$ respectively are in the same Teichmüller class. Furthermore, $|\mu_1(z)| = k$ for all $z \in G$ and $||\nu_1||_\infty \leq ||\mu_1||_\infty = k$. Now μ_1 is uniquely infinitesimally extremal, therefore we have $\widetilde{\mu_1}(z) = \widetilde{\nu_1}(z)$ for almost all $z \in h(f(G))$. Thus the quasiconformal mapping

$$(h \circ g)^{-1} \circ ((h \circ f)^{-1})^{-1} = g^{-1} \circ f$$

has dilatation equal to zero on G. Hence $\widetilde{\mu}(z) = \widetilde{\nu}(z)$ for almost all $z \in f(G)$, which completes the proof. □

Taking these two lemmas together gives what is called the Equivalence Theorem.

Theorem 9.5.2. *A Beltrami differential μ is uniquely extremal in its Teichmüller class if and only if it is uniquely infinitesimally extremal.*

9.6 Hahn–Banach extensions

To characterize unique extremality in general, we need to look at the proof of the Hahn–Banach theorem and find conditions on unique norm-preserving extensions of linear functionals. Let Y be a subspace of a normed space X and let Λ be a linear functional on X and λ its real part. Define

$$\bar{\lambda}(x_0) = \inf_{y \in X} \{\lambda(y) + ||\lambda|| ||y - x_0||\}$$

and

$$\underline{\lambda}(x_0) = \sup_{x \in X} \{\lambda(x) - ||\lambda|| ||x - x_0||\}.$$

Analysis of the proof of the Hahn–Banach yields the following lemma.

Lemma 9.6.1. *The linear functional Λ has a norm-preserving extension from X to Y if and only if $\bar{\lambda}(x_0) = \underline{\lambda}(x_0)$ for all $x_0 \in X \setminus Y$.*

We say that λ satisfies the unique approximation property at $x_0 \in X \setminus Y$ if there exists sequence u_n and v_n in Y such that

$$\lambda(u_n - v_n) = ||\lambda|| \left(||u_n - x_0|| + ||x_0 - v_n||\right) + o(1).$$

Define the functional $\delta = \delta_\lambda$ on Y by

$$\delta(\varphi) = ||\lambda|| ||\varphi|| - \lambda(\varphi).$$

The sequence φ_n in Y is called a weak Hamilton sequence for λ if $\delta_\lambda(\varphi_n)$ converges to zero as $n \to \infty$.

Lemma 9.6.2. *If λ has a unique norm-preserving extension from Y to X, then for all $x_0 \in X \setminus Y$, there exists sequence u_n and v_n in Y such that*

(a) $\lambda(u_n) = \lambda(x_0) + ||\lambda|| ||u_n - x_0|| + o(1)$,
(b) $\lambda(v_n) = \lambda(x_0) - ||\lambda|| ||v_n - x_0|| + o(1)$,
(c) $\lambda(u_n - v_n) = ||\lambda||(||x_0 - u_n|| + ||x_0 - v_n||) + o(1)$,
(d) *$u_n - v_n$ is a weak Hamilton sequence for λ.*

Proof. Without loss of generality we can assume that $||\lambda|| = 1$. If λ has a unique Hahn–Banach extension, which we also denote by λ, from Y to X, then by Lemma 9.6.1, for each $x_0 \in X \setminus Y$ there exist sequences u_n and v_n in X such that

$$\lambda(u_n) - ||u_n - x_0|| + o(1) = \lambda(x_0) = \lambda(v_n) + ||v_n - x_0|| + o(1).$$

Hence if $w_n = u_n - v_n$, then $\lambda(w_n) = ||x_0 - u_n|| + ||x_0 - v_n|| + o(1)$. Combining this with

$$\lambda(w_n) \le ||w_n|| \le ||x_0 - u_n|| + ||x_0 - v_n||,$$

we conclude that

$$||w_n|| = ||x_0 - u_n|| + ||x_0 - v_n|| + o(1)$$

and $\lambda(w_n) = ||w_n|| + o(1)$. □

As a consequence of the previous two lemmas, we have the following result.

Theorem 9.6.1. *Let Λ be a bounded linear functional on Y and λ its real part. Then Λ has a unique norm-preserving extension from Y to X if and only λ satisfies the unique approximation property at each $x_0 \in X \setminus Y$.*

Lemma 9.6.3. *If $\mu \in L^\infty(M)$ is uniquely infinitesimally extremal and $X(\mu)$ denotes the extremal set of μ, then there exists a sequence $\varphi_n \in A^1(M)$ such that $\delta(\varphi_n) \to 0$ and $\liminf |\varphi_n(z)| > 0$ for almost all $z \in X(\mu)$, where $\delta = \delta_\lambda$ corresponds to the linear functional λ, the real part of Λ_μ.*

Proof. Let μ be uniquely infinitesimally extremal and let $E = X(\mu)$ be its extremal set. Without loss of generality, we can assume that $||\mu||_\infty = 1$. Let $q \in A^1(M)$ have norm one, and let $f = |q|\bar{\mu}\chi_E$. Clearly $f \in L^1(M)$ and

$$\int_M f\mu = \int_E |q| = \int_E |f| = ||f||.$$

By condition (a) of Lemma 9.6.2, there exists a sequence $\varphi_n \in A^1(M)$ such that

$$\Re \int_M \mu f + ||f - \varphi_n|| - \Re \int_M \varphi_n \mu \to 0.$$

Hence

$$\begin{aligned} 0 &\le ||f|| + ||f - \varphi_n|| - ||\varphi_n|| \\ &\le \Re \int_M \mu f + ||f - \varphi_n|| - \Re \int_M \varphi_n \mu \to 0, \end{aligned}$$

and

$$\begin{aligned} 0 &\le ||\varphi_n|| - \Re \int_M \varphi_n \mu \\ &\le ||f|| + ||f - \varphi_n|| - \Re \int_M \varphi_n \mu \\ &= \Re \int_M \mu f + ||f - \varphi_n|| - \Re \int_M \varphi_n \mu \to 0. \end{aligned}$$

Thus $\delta(\varphi_n) \to 0$. Since

$$0 \le \int_E (|f| + |f - \varphi_n| - |\varphi_n|) \le ||f|| + ||f - \varphi_n|| - ||\varphi_n|| \to 0,$$

by passing to a subsequence we may assume that

$$|f(z)| + |f(z) - \varphi_n(z)| - |\varphi_n(z)| \to 0$$

for almost all $z \in E$. Hence $\liminf |\varphi_n(z)| \geq |f(z)| = |q(z)| > 0$ for almost all $z \in E$. □

Exercise 9.6.1. Prove the converse of this lemma. (Hint: first show that μ is infinitesimally extremal and then use the generalized Infinitesimal Delta Inequality (9.17).)

Putting together these results, we can generalize Theorem 9.4.1 and characterize unique extremality of an arbitrary Beltrami differential.

Theorem 9.6.2. *Let μ be a Beltrami differential in the unit ball of $L^\infty(M)$. Then the following conditions are equivalent:*

(a) *μ is uniquely extremal in its Teichmüller class,*

(b) *μ is uniquely infinitesimally extremal,*

(c) *for every admissible variation η of μ and every compact subset E of the extremal set $X(\mu)$ with non-zero measure, there exists a sequence $\varphi_n \in A^1(M)$, with $||\varphi_n|| = 1$ for all n, such that*

$$\frac{1}{\int_E |\varphi_n|}\left(||\eta||_\infty - \Re \int_M \eta\varphi_n\right) \to 0,$$

(d) *for every admissible variation η of μ and every compact subset E of $X(\mu)$ with non-zero measure, there exists a sequence $\varphi_n \in A^1(M)$, with $||\varphi_n|| = 1$ for all n, such that*

$$\frac{1}{\int_E |\varphi_n|}\Re \int_E \eta\varphi_n \to ||\eta||_\infty$$

and

$$\frac{1}{\int_E |\varphi_n|}\left(||\eta||_\infty \int_{M\setminus E} |\varphi_n| - \Re \int_{M\setminus E} \eta\varphi_n\right) \to 0,$$

(e) *μ is extremal in its Teichmüller class and for every $r > 0$, every admissible variation η of μ and every compact subset E of $X(\mu)$ with non-zero measure, $[\eta\chi_E + \frac{1}{1+r}\eta\chi_{M\setminus E}]$ is a Strebel point in $T(M)$,*

(f) *μ is infinitesimally extremal and for every $r > 0$, every admissible variation η of μ and every compact subset E of $X(\mu)$ with non-zero measure, $[\eta\chi_E + \frac{1}{1+r}\eta\chi_{M\setminus E}]$ corresponds to a Reich–Strebel functional,*

(g) *for every admissible variation η of μ there exists a sequence $\varphi_n \in A^1(M)$ such that $||\varphi_n|| ||\eta||_\infty - \Re \int_M \eta\varphi_n \to 0$ and $\liminf |\varphi_n(z)| > 0$ for almost all $z \in X(\mu)$,*

(h) *at each point of $L^1(M) \setminus A^1(M)$, λ_μ satisfies the unique approximation property.*

Proof. The equivalence of (a) and (b) is Theorem 9.5.2, and the equivalence of (b) and (h) follows from Theorem 9.6.1. The equivalence of (c) and (d) is left to the reader. If μ is any uniquely extremal Beltrami differential then so is every admissible variation η of μ, by the proof of Theorem 9.5.2. Thus by Lemma 9.6.3, (b) implies (g). Again, if μ satisfies (b), so does any admissible variation η of μ, thus by the proof of Lemma 9.4.2, (b) implies (f). Using the proof of Lemma 9.4.1 shows that (a) implies (e). Using the proofs of Lemmas 9.4.1 and 9.4.2 shows that (e) implies (c) and (f) implies (c) respectively. From the generalized Infinitesimal Delta Inequality of Theorem 9.3.2, (c) or (g) implies that every admissible variation η of μ is uniquely extremal on its extremal set $X(\mu)$, and so by the proof of Theorem 9.5.2, (c) implies (b) and (g) implies (b). □

10

Isomorphisms of Teichmüller space

In this chapter, we show how the Kobayashi metric is equal to the Teichmüller metric on $T(M)$ and investigate how this relates to biholomorphic maps between Teichmüller spaces. We will classify all biholomorphic maps between two Teichmüller spaces by reducing the problem to the cotangent space. That is, all surjective linear isometries between Bergman spaces of Riemann surfaces of non-exceptional type are geometric, which in particular implies that the two Riemann surfaces are conformally related.

10.1 The Kobayashi metric

Recall that the space of Beltrami differentials, $B(M)$, is the open unit ball in $L^\infty(M)$. The Teichmüller metric on $B(M)$ is defined by

$$\mathrm{d}_{B(M)} = \tanh^{-1} \left|\left| \frac{\mu - \nu}{1 - \overline{\mu}\nu} \right|\right|_\infty ,$$

for all $\mu, \nu \in B(M)$. The right-hand side of this expression equals the essential supremum of $\{\rho_{\mathbb{D}}(\mu(t), \nu(t)) : z \in \mathbb{H}\}$, where $\rho_{\mathbb{D}}$ is the hyperbolic metric on $\mathbb{H}$.

Now let X be any connected complex Banach manifold, and let $H(\mathbb{D}, X)$ be the set of all holomorphic maps from $\mathbb{D}$ into X. The Kobayashi function $\delta_X : X \times X \to [0, +\infty]$ is

$$\delta_X(x, y) = \inf \{\rho_{\mathbb{D}}(0, t) : f(0) = x, f(t) = y \text{ for some } f \in H(\mathbb{D}, X)\} ,$$

provided the set of such maps is non-empty, and $+\infty$ otherwise. If X and Y are connected complex Banach manifolds and $f : X \to Y$ is holomorphic, then

$$\delta_Y(f(x_1), f(x_2)) \leq \delta_X(x_1, x_2),$$

for all $x_1, x_2 \in X$, and with equality if f is biholomorphic.

Definition 10.1.1. *The Kobayashi pseudo-metric σ_X on X is the largest pseudo-metric on X such that*

$$\sigma_X(x,y) \le \delta_X(x,y), \tag{10.1}$$

for all $x, y \in X$. If δ_X is a metric, as it will be in our cases, then σ_X is a metric and σ_X and δ_X are equal (see [28]).

Note that when we say σ_X is the largest pseudo-metric, we mean that if d_X is any other pseudo-metric which satisfies (10.1) with d_X in place of σ_X, then

$$d_X(x,y) \le \sigma_X(x,y)$$

for all $x, y \in X$.

Example 10.1.1. If M is $\mathbb{C}$ or $\widehat{\mathbb{C}}$, then the linear maps $f_n(z) = nz$, together with their inverses, from $\mathbb{D}$ into M show that $\delta_X \equiv 0$. Hence $\sigma_M \equiv 0$. Furthermore, if M is a torus or the punctured plane $\mathbb{C} \setminus \{0\}$, then there is a covering map π from $\mathbb{C}$ onto M. Since π is a contraction in the respective pseudo-metrics, $\sigma_M \equiv 0$.

Example 10.1.2. Let M be a hyperbolic Riemann surface, that is, it has $\mathbb{D}$ as its universal cover. If π is the covering map from $\mathbb{D}$ onto M, then $\pi \in H(\mathbb{D}, M)$ and so is a contraction in the corresponding Kobayashi pseudo-metrics. However, every $f \in H(\mathbb{D}, M)$ lifts to a map $\tilde{f} \in H(\mathbb{D}, \mathbb{D})$ such that $f = \pi \circ g$. Thus δ_X is equal to the quotient pseudo-metric on M with respect to the covering map π and the hyperbolic metric on $\mathbb{D}$. That is, δ_X coincides with the hyperbolic metric on M. Since δ_X is a pseudo-metric, $\sigma_M = \delta_X$ and so the Kobayashi pseudo-metric is equal to the hyperbolic metric on M.

Example 10.1.3. Let B be the unit ball in a complex Banach manifold X and pick $x \in B$. The linear function $f(t) = tx/||x||$ maps the unit disk $\mathbb{D}$ into the unit ball B and maps $||x||$ to x and 0 to 0. Therefore

$$\sigma_B(0,x) \le \sigma_{\mathbb{D}}(0, ||x||).$$

However, via the Hahn–Banach theorem, there exists a continuous linear functional L on X such that $L(x) = ||x||$ and $||L|| = 1$. Thus, L maps B into the unit disk $\mathbb{D}$ and so

$$\sigma_{\mathbb{D}}(0, ||x||) \le \sigma_B(0,x).$$

The definition of the hyperbolic metric on $\mathbb{D}$ leads to

$$\sigma_B(0,x) = \frac{1}{2}\tanh^{-1}||x||. \tag{10.2}$$

Proposition 10.1.1. *For all $\mu, \nu \in B(M)$, we have*

$$\delta_{B(M)}(\mu,\nu) = \mathrm{d}_{B(M)}(\mu,\nu).$$

Proof. We can assume that $\mu \neq \nu$. First, we also assume that $\mu = 0$. Suppose that $f \in H(\mathbb{D}, B(M))$, $f(0) = 0$ and $f(t) = \nu$ for some $t \in \mathbb{D}$. By the Schwarz lemma,

$$|t| \geq ||f(t)||_\infty = ||\nu||_\infty.$$

Taking the infimum over all such f,

$$\delta_{B(M)}(0,\nu) \geq \rho_{\mathbb{D}}(0, ||\nu||_\infty = \mathrm{d}_{B(M)}(0,\nu).$$

Choose the function $f(t) = t\nu/||\nu||_\infty$ and observe that

$$\delta_{B(M)}(0,\nu) \leq \rho_{\mathbb{D}}(0, ||\nu||_\infty) = \mathrm{d}_{B(M)}(0,\nu)$$

since $f(||\nu||_\infty) = \nu$. Now if $\mu \neq 0$, observe that the function $f : B(M) \to L^\infty(M)$ defined by

$$f(\lambda) = \frac{\mu - \lambda}{1 - \overline{\mu}\lambda}$$

is a holomorphic map of $B(M)$ onto itself. Therefore

$$\delta_{B(M)}(\mu,\nu) = \delta_{B(M)}(f(\mu), f(\nu)) = \mathrm{d}_{B(M)}(0, f(\nu)) = \mathrm{d}_{B(M)}(\mu,\nu).$$

□

As we have seen, the Teichmüller metric on $B(M)$ induces a quotient metric on $T(M)$:

$$\mathrm{d}_T([\mu],[\nu]) = \inf\left\{\mathrm{d}_{B(M)}(\mu_0,\nu_0) : \mu_0,\nu_0 \in B(M), \mu_0 \in [\mu], \nu_0 \in [\nu]\right\},$$

for all $[\mu],[\nu] \in T(M)$. We can now prove Royden's theorem on the equality of Teichmüller and Kobayashi metrics on $T(M)$, proved in [45].

Theorem 10.1.1. *The Teichmüller and Kobayashi metrics on $T(M)$ coincide, that is,*

$$\mathrm{d}_T([\mu],[\nu]) = \delta_{T(M)}([\mu],[\nu]).$$

Proof. Fix $[\mu],[\nu] \in T(M)$. We have

$$\delta_{T(M)}([\mu],[\nu]) \leq \inf\left\{\delta_{B(M)}(\mu_0,\nu_0) : \mu_0 \in [\mu], \nu_0 \in [\nu]\right\}.$$

Hence by Proposition 10.1.1,

$$\delta_{T(M)}([\mu],[\nu]) \leq \mathrm{d}_T([\mu],[\nu]).$$

For the opposite inequality, choose $f \in H(\mathbb{D}, T(M))$ such that $f(0) = [\mu]$ and $f(t) = [\nu]$ for some $t \in \mathbb{D}$. Use Theorem 7.7.4 to write $\pi \circ g = f$ with

$$g \in H(\mathbb{D}, B(M)).$$

Using Proposition 10.1.1,

$$\rho_{\mathbb{D}}(0,t) \geq \mathrm{d}_{B(M)}(g(0), g(t)) \geq \mathrm{d}_T(\pi(g(0)), \pi(g(t))) = \mathrm{d}_T([\mu],[\nu]).$$

Taking the infimum over all such f, we obtain

$$\delta_{T(M)}([\mu],[\nu]) \geq \mathrm{d}_T([\mu],[\nu]).$$

□

10.2 Equimeasurability

Let $C_0(\mathbb{R}^n)$ be the space of all continuous complex functions on $\mathbb{R}^n$ that vanish at infinity. In this section, the Fourier transform of $f \in L^1(\mathbb{R}^n)$ will be denoted by $\widetilde{f}$, and $f \star g$ will denote the convolution of the functions f and g. If u is a function defined on $\mathbb{R}^n$, then the a-translate of u, for $a \in \mathbb{R}^n$ is

$$(\tau_a u)(x) = u(x+a),$$

for $x \in \mathbb{R}^n$ and the t-dilation u_t of u, for $0 < t < \infty$, is

$$u_t(x) = u(tx),$$

for $x \in \mathbb{R}^n$. The function u is called radial if $u \circ T = u$ for every rotation T of $\mathbb{R}^n$.

Lemma 10.2.1. *Let $u \in C_0(\mathbb{R}^n)$ be radial and not identically zero. Let V be the smallest closed translation invariant subspace of $C_0(\mathbb{R}^n)$ that contains all dilations u_t of u. Then $V = C_0(\mathbb{R}^n)$.*

Proof. Let W be the space of all $f \in L^1(\mathbb{R}^n)$ such that $f \star u_t = 0$ for all t. If $V \neq C_0$, then there exists a complex measure $\mu \neq 0$ such that $u_t \star \mu = 0$ for all t, and there exists $g \in L^1$ such that $g \star \mu \neq 0$. Since $g \star \mu \in W$, $W \neq \{0\}$. Therefore it is enough to show that $W = \{0\}$.

Let $\widetilde{W} = \{\widetilde{f} : f \in W\}$, and let Z be the intersection of the zero sets of all $\widetilde{f} \in \widetilde{W}$. Now, if $f \in L^1(\mathbb{R}^n)$, $u \in C_0(\mathbb{R}^n)$ and T is an invertible linear transformation on $\mathbb{R}^n$, then we have the following two standard formulas,

$$|\det T|[f \star (u \circ T)](x) = [(f \circ T^{-1}) \star u](Tx)$$

and

$$|\det T|\widetilde{(f \circ T)}(x) = \widetilde{f}(Sx),$$

where S is the transpose of T^{-1}. Using these two formulas, we see that W and $\widetilde{W}$ are dilation invariant. If T is a rotation, then $u \circ T = u$, since u is radial, and therefore W and $\widetilde{W}$ are rotation invariant. Therefore Z is a closed subset of $\mathbb{R}^n$ which is invariant under dilations and rotations, which implies Z is either $\emptyset$, $\{0\}$ or $\mathbb{R}^n$.

Note that W is a closed translation invariant subspace of L^1. If $Z = \emptyset$, then Wiener's theorem implies that $W = L^1$ and hence that $V = \{0\}$, which is a contradiction.

If $Z = \{0\}$, then a slight variation on Wiener's theorem shows that $f \in W$ if and only if $\widetilde{f}(0) = 0$. If $g \in L^1$ and $a \in \mathbb{R}^n$, then $g - \tau_a g \in W$. Then

$$(u - \tau_a u) \star g = u \star (g - \tau_a g) = 0.$$

This implies that u is a constant, and hence identically zero, which is again a contradiction. Therefore Z must be $\mathbb{R}^n$ and $W = \{0\}$. □

If $z = (z_1, \ldots, z_n)$ and $w = (w_1, \ldots, w_n)$ are elements of $\mathbb{C}^n$, then the inner product

$$< z, w > = z_1\overline{w_1} + \cdots + z_n\overline{w_n}$$

turns $\mathbb{C}^n$ into a Hilbert space with norm

$$|z| = < z, z >^{1/2}.$$

The open unit ball B_n of $\mathbb{C}^n$ is the set of all $z \in \mathbb{C}^n$ with $|z| < 1$. The boundary S of B_n is a sphere of real dimension $2n - 1$, which carries a unique rotation-invariant Borel probability measure σ. For brevity we omit any mention of n in S and σ.

Lemma 10.2.2. *If $f : \mathbb{C} \to \mathbb{C}$ is continuous, $n > 1$, and if*

$$\psi(z) = \int Sf(< z, w >)\mathrm{d}\sigma(w), \tag{10.3}$$

for $z \in \mathbb{C}^n$, then

$$\psi(z) = \frac{n-1}{\pi} \int_0^1 (1 - r^2)^{n-2} r \,\mathrm{d}r \int_{-\pi}^{\pi} f(|z|r\mathrm{e}^{i\theta})\,\mathrm{d}\theta. \tag{10.4}$$

In particular, ψ is radial.

Proof. The proof is a calculation which we sketch.

Fix z and choose coordinates in $\mathbb{C}^n$ such that via a unitary transformation, $z = (|z|, 0, \ldots, 0)$. Then we have $< z, w >= |z|\overline{w_1}$ and so the integrand in (10.3) depends only on the first coordinate of w. Thus (10.3) can be transformed to an integral over the unit disk in $\mathbb{C}$ and, via a calculation, yields (10.4). □

Lemma 10.2.3. *Let $0 < p < \infty$, $p \neq 2, 4, 6 \ldots$ and suppose $0 < t_1 < \cdots < t_N$ where $2N > p + 2$. Then there exists real numbers $a_1, \ldots, a_N$ such that if $n \geq 1$,*

$$\psi_n(z) = \int_S |1+ < z, w > |^p \,\mathrm{d}\sigma(w), \tag{10.5}$$

for $z \in \mathbb{C}^n$, and if

$$u_n(z) = \sum_{i=1}^{N} a_i \psi_n(t_i z), \tag{10.6}$$

for $z \in \mathbb{C}^n$, then u_n vanishes at infinity without being identically zero. Further, u_n is continuous and radial.

Proof. When $n = 1$, S is the unit circle and so

$$\psi_1(\lambda) = \frac{1}{2\pi} \int_{-\pi}^{\pi} |1 + \lambda \mathrm{e}^{-i\theta}|^p \,\mathrm{d}\theta, \tag{10.7}$$

for $\lambda \in \mathbb{C}$. When $n > 1$, Lemma 10.2.2 implies

$$\psi_n(z) = 2(n-1) \int_0^1 \psi_1(r|z|)(1 - r^2)^{n-2} r \,\mathrm{d}r, \tag{10.8}$$

for $z \in \mathbb{C}^n$. Hence

$$u_n(z) = 2(n-1)\int_0^1 u_1(r|z|)(1-r^2)^{n-2} r\,\mathrm{d}r \tag{10.9}$$

no matter how $a_1, \ldots, a_N$ are chosen. Therefore u_n is continuous and radial, and the most important case to prove is the $n = 1$ case. First,

$$(1+\lambda)^{p/2} = \sum_{k=0}^{\infty} b_k \lambda^k,$$

for $|\lambda| < 1$. Since $p/2 \notin \mathbb{Z}$, the binomial coefficients b_k are non-zero for all k. The integrand in (10.7) is then

$$\left| \sum_{k=0}^{\infty} b_k \lambda^k \mathrm{e}^{-ik\theta} \right|^2.$$

Hence

$$\psi_1(\lambda) = \sum_{k=0}^{\infty} b_k^2 |\lambda|^{2k},$$

for $|\lambda| < 1$. However $a_1, \ldots, a_N$ are chosen,

$$u_1(\lambda) = \sum_{k=0}^{\infty} b_k^2 c_k |\lambda|^{2k},$$

for $|\lambda| < t_N^{-1}$, where

$$c_k = \sum_{i=1}^{N} a_i t_i^{2k},$$

for $k = 0, 1, 2, \ldots$. If not all the a_i are zero, then, since the t_i are distinct, c_k can be zero for at most finitely many k. Thus u_1 is not identically zero. Moreover, inserting the power series for u_1 into (10.9) and integrating term-by-term (which is valid for $|z| < t_N^{-1}$), u_n does not vanish identically for any $n > 1$.

Now, (10.7) gives the functional equation

$$\psi_1(\lambda) = |\lambda|^p \psi_1\left(\frac{1}{\lambda}\right),$$

for $0 < |\lambda| < \infty$. Hence

$$\psi_1(\lambda) = \sum_{k=0}^{\infty} b_k^2 |\lambda|^{p-2k},$$

for $|\lambda| > 1$. The set of $N-1$ homogeneous linear equations

$$\sum_{i=1}^{N} t_i^{p-2k} a_i = 0$$

for $k = 0, \ldots, N-2$, has a non-trivial solution $(a_1, \ldots, a_N)$, which is used to define u_n. As we saw earlier, u_n is not identically zero. Thus

$$u_1(\lambda) = \sum_{k=0}^{\infty} b_k^2 \gamma_k |\lambda|^{p-2k},$$

for $|\lambda| > t_1^{-1}$, where

$$\gamma_k = \sum_{i=1}^{N} t_i^{p-2k} a_i,$$

for $k = 0, 1, 2, \ldots$. The choice of the a_i shows that the first $N-1$ terms in the series for u_1 are zero. Thus

$$u_1(\lambda) = O(|\lambda|^{p-2N+2})$$

as $|\lambda| \to \infty$. Since $2N > p+2$, $u_1(\lambda) \to 0$ as $|\lambda| \to \infty$. By the dominated convergence theorem, $u_n(z) \to 0$ as $|z| \to \infty$, if $n > 1$, which completes the proof. $\square$

Definition 10.2.1. *Suppose that $f_1, \ldots, f_n$ are μ-measurable functions on X and that $g_1, \ldots, g_n$ are ν-measurable functions on Y. Writing $F = (f_1, \ldots, f_n)$ and $G = (g_1, \ldots, g_n)$, which we consider as $\mathbb{C}^n$ valued functions, then F and G are equimeasurable if*

$$\mu(F^{-1}(E)) = \nu(G^{-1}(E))$$

for every Borel set $E \subset \mathbb{C}$.

The following theorem on a condition for equimeasurability and the previous lemmas are due to Rudin in [46].

Theorem 10.2.1. *Let $0 < p < \infty$, $p \neq 2, 4, 6, \ldots$, $n \in \mathbb{N}$, and let μ and ν be measures on measurable spaces X and Y respectively. If for $1 \leq i \leq n$, $f_i \in L^p(\mu)$ and $g_i \in L^p(\nu)$, and*

$$\int_X |1 + z_1 f_1 + \cdots + z_n f_n|^p \, d\mu = \int_Y |1 + z_1 g_1 + \cdots z_n g_n|^p \, d\nu \tag{10.10}$$

for all $(z_1, \ldots, z_n) \in \mathbb{C}^n$, then $(f_1, \ldots, f_n)$ and $(g_1, \ldots, g_n)$ are equimeasurable.

Proof. Fix $0 < p < \infty$ with $p \neq 2, 4, 6, \ldots$ and let $n \in \mathbb{N}$. Write $F = (f_1, \ldots, f_n)$ and $G = (g_1, \ldots, g_n)$ where $f_i \in L^p(\mu)$ and $g_i \in L^p(\nu)$ for $i = 1, \ldots, n$. The hypotheses of the theorem can be written as

$$\int_X |1 + < F, w > |^p \, d\mu = \int_Y |1 + < G, w > |^p \, d\nu, \tag{10.11}$$

for all $w \in \mathbb{C}^n$. We abbreviate these hypotheses by writing $F \sim G$. If $F \sim G$, $a \in \mathbb{C}^n$, and $0 < t < \infty$, then it is easy to see that $F + a \sim G + a$ and $tF \sim tG$.

Now let

$$V = \left\{ h \in C_0(\mathbb{R}^{2n}) : \int_X (h \circ F)\, d\mu = \int_Y (h \circ G)\, d\nu \text{ whenever } F \sim G \right\}.$$

Then V is invariant under translations and dilations, and is clearly a closed subspace of $C_0(\mathbb{R}^{2n})$. We will show that V contains a non-trivial radial function. Define ψ_n as in (10.5). Since the left-hand side of (10.11) is continuous in w,

$$\int_S d\sigma(w) \int_X |1+ < F, w > |^p \, d\mu = \int_X (\psi_n \circ F)\, d\mu, \tag{10.12}$$

so that $\psi_n \circ F \in L^1(\mu)$. The left-hand side of (10.12) is unchanged if F and μ are replaced by G and ν respectively. Hence

$$\int_X (\psi_n \circ F)\, d\mu = \int_Y (\psi_n \circ G)\, d\nu.$$

This hold for all dilations of ψ_n, and hence it holds with u_n in place of ψ_n, where u_n is defined as in (10.6). By Lemma 10.2.3, $u_n \in C_0(\mathbb{R}^{2n})$ and thus $u_n \in V$. Now Lemma 10.2.1 implies that $V = C_0(\mathbb{R}^{2n})$.

Now the indicator function $\chi_E \in V$ for all compact sets E. Using the fact that $\{E : \chi_E \in V\}$ is a σ-algebra, and that $F \sim G$ is equivalent to saying V contains χ_E for every Borel set $E \subset \mathbb{C}^n$ completes the proof. □

10.3 Isometries of Bergman spaces

In this section, we will prove that an isometry between the Bergman spaces of two Riemann surfaces implies that the Riemann surfaces themselves are biholomorphically equivalent, when the Riemann surfaces are of non-exceptional type. In this section, we will prove the case where the Riemann surfaces are of finite analytic type.

Let $\widetilde{M}$ be a compact Riemann surface of genus $g \geq 0$ and let E be a finite, possibly empty, subset of $\widetilde{M}$ which contains exactly $n \geq 0$ points. We assume that $2g + n \geq 5$, so that the Riemann surface $M = \widetilde{M} \setminus E$ has non-exceptional finite type.

We discuss projective embeddings of $\widetilde{M}$ associated with $A^1(M)$. Let k be a positive integer, and let $\mathbb{P}^k$ be the k-dimensional complex projective space. Each point $(z_0, \ldots, z_k) \in \mathbb{C}^{k+1} \setminus \{0\}$ determines a point $[(z_0, \ldots, z_k)] \in \mathbb{P}^k$. The formula

$$\pi_0(z_1, \ldots, z_k) = [(1, z_1, \ldots, z_k)]$$

defines a holomorphic map of $\mathbb{C}^k$ onto a dense open subset of $\mathbb{P}^k$.

Proposition 10.3.1. *Let $M = \widetilde{M} \setminus E$ as above, and let $\varphi_0, \ldots, \varphi_k$ be a basis for $A^1(M)$. Set $f_j = \varphi_j / \varphi_0$, for $j = 1, \ldots, k$ and set*

$$M_0 = \widetilde{M} \setminus \{x \in \widetilde{M} : \text{ some } f_j \text{ has a pole at } x\}.$$

Let $F : M_0 \to \mathbb{C}^k$ be the holomorphic map $F = (f_1, \ldots, f_k)$. There is a unique holomorphic embedding $\Phi : \widetilde{M} \to \mathbb{P}^k$ such that

$$\Phi(x) = \pi_0(F(x)),$$

for all $x \in M_0$.

Proof. Consider the divisor $D = (\varphi_0) + \chi_E$ on $\widetilde{M}$. Clearly, $A^1(M)$ is the set of meromorphic quadratic differentials $\varphi = f\varphi_0$ such that $f \in \mathcal{O}_D(\widetilde{M})$.

Again, $\deg(D) \geq 2g + 1$. Since the functions $1, f_1, \ldots, f_k$ are a basis for $\mathcal{O}_D(\widetilde{M})$, we can conclude the map

$$x \mapsto [(1, f_1(x), \ldots, f_k(x))],$$

for $x \in \widetilde{M}$, when interpreted appropriately at the poles of the f_j, defines a holomorphic embedding of $\widetilde{M}$ into $\mathbb{P}^k$. □

Corollary 10.3.1. *The map F defined above is a homeomorphism of M_0 onto a closed subset of $\mathbb{C}^k$.*

Proof. Since F is holomorphic on M_0, it is continuous. Since $\Phi^{-1} \circ \pi_0 \circ F$ is the identity map of M_0 to itself, F is a homeomorphism. To see that $F(M_0)$ is a closed set, consider a sequence (x_n) in M_0 such that $F(x_n)$ converges to $z = (z_1, \ldots, z_k)$ in $\mathbb{C}^k$. We may assume that x_n converges to some point $\widetilde{x} \in \widetilde{M}$. Then $f_j(\widetilde{x}) = z_j \neq \infty$ for $j = 1, \ldots, k$ and so $\widetilde{x} \in M_0$ and $z = F(\widetilde{x})$. □

Definition 10.3.1. *If M and N are Riemann surfaces, then a surjective linear isometry $T : A^1(M) \to A^1(N)$ is called geometric if there exists a conformal map $\alpha : M \to N$ and a complex number θ with $|\theta| = 1$ such that*

$$T^{-1}(\varphi) = \theta(\varphi \circ \alpha)\alpha'^2,$$

for every $\varphi \in A^1(N)$.

The following theorem is proved in [17].

Theorem 10.3.1. *Suppose that M and N are Riemann surfaces of non-exceptional finite type and that $T : A^1(M) \to A^1(N)$ is a surjective complex-linear isometry. Then T is geometric.*

Proof. We will prove this theorem in the case where the given Riemann surfaces M and N are the complements of finite sets in compact Riemann surfaces $\widetilde{M}$ and $\widetilde{N}$. Note that we do not assume $\widetilde{M}$ and $\widetilde{N}$ have the same genus.

Let $\varphi_0, \ldots, \varphi_k$ be a basis for $A^1(M)$, and define M_0 and the map

$$F = (f_1, \ldots, f_k)$$

as in the previous proposition. Set $\psi_j = T(\varphi_j)$ for $j = 0, \ldots, k$. Since

$$T : A^1(M) \to A^1(N)$$

is a surjective $\mathbb{C}$-linear isometry, $\psi_0, \ldots, \psi_k$ is a basis for $A^1(N)$. Set $g_j = \psi_j/\psi_0$, for $j = 1, \ldots, k$, and set

$$N_0 = \widetilde{N} \setminus \{y \in \widetilde{N} : \text{ some } g_j \text{ has a pole at } y\}.$$

Let $G : N_0 \to \mathbb{C}^k$ be the holomorphic embedding map $G = (g_1, \ldots, g_k)$. By Proposition 10.3.1, there is a holomorphic embedding $\Psi : \widetilde{N} \to \mathbb{P}^k$ such that $\Psi = \pi_0 \circ G$ on N_0.

Let μ and ν be the finite positive Borel measures on M_0 and N_0 defined by

$$\mu(A) = \int_A |\varphi_0|$$

for all Borel sets $A \subset M_0$, and

$$\nu(B) = \int_B |\psi_0|$$

for all Borel sets $B \subset N_0$. Since T is a $\mathbb{C}$-linear isometry, we have

$$\int_{M_0} \left| 1 + \sum_{j=1}^k \lambda_j f_j \right| \mathrm{d}\mu = \int_M \left| \varphi_0 + \sum_{j=1}^k \lambda_j \varphi_j \right|$$

$$= \int_N \left| \psi_0 + \sum_{j=1}^k \lambda_j \psi_j \right| = \int_{N_0} \left| 1 + \sum_{j=1}^k \lambda_j g_j \right| \mathrm{d}\nu,$$

for all $(\lambda_1, \ldots, \lambda_k) \in \mathbb{C}^k$. Therefore, the maps F and G, and the measures μ and ν satisfy Rudin's equimeasurability condition, Theorem 10.2.1. Applying this to the closed subset $F(M_0)$ of $\mathbb{C}^k$, we obtain

$$||\varphi_0|| = \int_{M_0} |\varphi_0| = \mu(M_0) = \nu(G^{-1}(F(M_0)))$$

$$= \int_{G^{-1}(F(M_0))} |\psi_0| \leq \int_{N_0} |\psi_0| = ||\psi_0||.$$

Since $||\psi_0|| = ||\varphi_0||$, the weak inequality here is actually an equality, and then $G^{-1}(F(M_0))$ has full measure in N_0. Since it is a closed subset of N_0, $G^{-1}(F(M_0))$ equals N_0, and $G(N_0)$ is contained in $F(M_0)$.

Similarly, applying the equimeasurability condition of Theorem 10.2.1 to the set $G(N_0)$, we find that $F(M_0)$ is a subset of $G(N_0)$. Therefore the sets $F(M_0)$ and $G(N_0)$ are equal, and so are their images under the map π_0 from $\mathbb{C}^k$ to $\mathbb{P}^k$. Now $\pi_0(F(M_0)) = \Phi(M_0)$ is dense in the compact set $\Phi(\widetilde{M})$, and $\pi_0(G(N_0))$ is dense in the compact set $\Psi(\widetilde{N})$, and so the sets $\Phi(\widetilde{M})$ and $\Psi(\widetilde{N})$ are equal.

Let $h : \widetilde{N} \to \widetilde{M}$ be the bijective holomorphic map $\Phi^{-1} \circ \Psi$. The restriction of h to N_0 satisfies $F \circ h = G$ and $h(N_0) = M_0$. From the definitions of F and G, we obtain

$$\frac{T(\varphi_j)}{T(\varphi_0)} = \frac{\psi_j}{\psi_0} = g_j = f_j \circ h = \frac{\varphi_j}{\varphi_0} \circ h = \frac{h^*(\varphi_j)}{h^*(\varphi_0)}$$

for $j = 1, \ldots, k$, and so

$$\frac{T(\varphi)}{T(\varphi_0)} = \frac{h^*(\varphi)}{h^*(\varphi_0)}$$

for all $\varphi \in A^1(M)$, where we write $h^*(\varphi)$ for the pullback of φ by h, $(\varphi \circ h)h'^2$. Let K be any compact set in N_0. Applying the equimeasurability condition to the compact set $G(K)$ in $\mathbb{C}^k$, we obtain

$$\int_K |T(\varphi_0)| = \int_K |\psi_0| = \nu(K) = \mu(h(K)) = \int_{h(K)} |\varphi_0| = \int_K |h^*(\varphi_0)|.$$

Since K is arbitrary, we must have $|T(\varphi_0)| = |h^*(\varphi_0)|$ in N_0, and hence in all of $\widetilde{N}$. Therefore $T(\varphi_0) = \mathrm{e}^{it} h^*(\varphi_0)$ for some $t \in \mathbb{R}$, and we see that

$$T(\varphi) = \mathrm{e}^{it} h^*(\varphi), \tag{10.13}$$

for all $\varphi \in A^1(M)$. To complete the proof, we need to show that $h(N) = M$. By an earlier proposition, N is the set of points in $\widetilde{N}$ where every $T(\varphi)$ is finite, and $h^{-1}(M)$ is the set of points in $\widetilde{N}$ where every $h^*(\varphi)$ is finite. These sets coincide by (10.13). □

10.4 Geometric isometries in the general case

In this section, we outline how to prove Theorem 10.3.1 when we do not restrict attention to the finite analytic case (see [35]).

Definition 10.4.1. *The subset S_M of a Riemann surface M is defined as follows: $p_0 \in S_M$ if for every point $p \in M$ with $p \neq p_0$, there exists $\phi \in A^1(M)$ such that*

$$\phi(p)\phi(p_0) = 0$$

and such that $\phi(p)$ and $\phi(p_0)$ are not both equal to zero.

Note that although $\phi \in A^1(M)$ is not a function on M, the statements that $\phi(p) = 0$ and $\phi(p_0) = 0$ are well defined, and so the above definition is independent of choice of local parameter. The following theorem we state without proof, which can be found in [35].

Theorem 10.4.1. *Suppose that M is a Riemann surface that is not of exceptional type. Then the set $M \setminus S_M$ is a discrete set of points in M. Furthermore, if p' is a puncture in the boundary of M, then p' is not a cluster point of the set $M \setminus S_M$.*

By saying that p' is a puncture in the boundary of M, we mean that there exists another Riemann surface $\widetilde{M}$ such that the surface M is obtained from $\widetilde{M}$ by deleting the point p'.

We now assume that M and N are Riemann surfaces that are not of exceptional type. Let $T : A^1(M) \to A^1(N)$ be a surjective linear isometry. Since

$A^1(M)$ is a separable Banach space, we can choose a sequence of distinct differentials $\{\phi_k\}$, $k = 0, 1, 2, \ldots$, such that for every $\phi \in A^1(M)$, there exists a subsequence of $\{\phi_k\}$ which is norm converging to ϕ. We also choose ϕ_0 and ϕ_1 to be linearly independent.

Let S_1 be the discrete set of points where ϕ_0 has zeros, and let S_2 be the discrete set of points where the meromorphic function ϕ_1/ϕ_0 has first derivative equal to zero. Denote by M_1 the set $S_M \setminus (S_1 \cup S_2)$. If p' is a puncture in the boundary of M, then ϕ_1/ϕ_0 is a well-defined meromorphic function in a given neighbourhood of p'. Therefore S_1 and S_2 does not have the point p' as a cluster point. Therefore $M \setminus M_1$ does not have the point p' as a cluster point. We define the set N_1 in N in the same way, using the sequence $\{T(\phi_k)\}$ in $A^1(N)$. We will for the moment work with M_1 and N_1 and return to M and N later.

For every $n \in \mathbb{N}$, we define the maps $I_n : M_1 \to \mathbb{C}^n$ and $J_n : N_1 \to \mathbb{C}^n$ by

$$I_n(p) = \left(\frac{\phi_1}{\phi_0}(p), \ldots, \frac{\phi_n}{\phi_0}(p)\right),$$

for $p \in M_1$, and

$$J_n(q) = \left(\frac{T(\phi_1)}{T(\phi_0)}(q), \ldots, \frac{T(\phi_n)}{T(\phi_0)}(q)\right),$$

for $q \in N_1$. The maps I_n and J_n are holomorphic and, in particular, from the definition of M_1, it follows that for every $p \in M_1$, there is a neighbourhood U of p, and a biholomorphic map $\widetilde{I_n} : U \times \mathbb{C}^{n-1} \to \mathbb{C}^n$, such that $I_n(p) = \widetilde{I_n}(p, 0)$. This implies that if F is a Borel set on M_1, then $I_n(F)$ is a Borel set in $\mathbb{C}^n$. An analogous observation holds for N_1.

Also define the maps $I_\infty : M_1 \to \mathbb{C}^{\mathbb{N}}$ and $J_\infty : N_1 \to \mathbb{C}^{\mathbb{N}}$ as

$$I_\infty(p) = \left(\frac{\phi_1}{\phi_0}(p), \ldots\right),$$

for $p \in M_1$, and

$$J_\infty(q) = \left(\frac{T(\phi_1)}{T(\phi_0)}(q), \ldots\right),$$

for $q \in N_1$. The product topology of pointwise convergence makes the space $\mathbb{C}^{\mathbb{N}}$ a Fréchet space. Then the maps I_∞ and J_∞ are holomorphic, and they map Borel sets to Borel sets for the same reason that I_n and J_n do.

Suppose that $z = x + iy$ and $z_1 = x_1 + iy_1$ are local parameters on M and N respectively. Set $\mathrm{d}\mu = |\phi_0| \,\mathrm{d}x\,\mathrm{d}y$, and $\mathrm{d}\nu = |T(\phi_0)| \,\mathrm{d}x_1\,\mathrm{d}y_1$. The measures μ and ν are well defined on M and N respectively.

Lemma 10.4.1. *Let $p \in M_1$ and $q \in N_1$. Then*

$$\bigcap_{n=1}^{\infty} I_n^{-1}(I_n(p)) = \{p\},$$

and

$$\bigcap_{n=1}^{\infty} J_n^{-1}(J_n(q)) = \{q\}.$$

That is, I_∞ and J_∞ are injective.

Proof. Suppose that these exists $p' \in M$ with $p \neq p'$ such that

$$p' \in \bigcap_{n=1}^{\infty} I_n^{-1}(I_n(p)).$$

Since ϕ_n is dense in $A^1(M)$, we conclude that

$$\frac{\phi}{\phi_0}(p) = \frac{\phi}{\phi_0}(p') \tag{10.14}$$

for every $\phi \in A^1(M)$. From Theorem 10.4.1, we know that there exists $\psi \in A^1(M)$ such that either $\psi(p) = 0$ and $\psi(p') \neq 0$, or $\psi(p) \neq 0$ and $\psi(p') = 0$. This contradicts (10.14). The proof for N_1 is similar. □

Let G_M be the set of all points $p \in M_1$ such that the set

$$\bigcap_{n=1}^{\infty} J_n^{-1}(I_n(p))$$

is a non-empty subset of N_1. Also, let G_N be the set of all points $q \in N_1$ such that the set

$$\bigcap_{n=1}^{\infty} I_n^{-1}(J_n(q))$$

is a non-empty set of M_1. Observe that

$$G_M = I_\infty^{-1}(J_\infty(N_1)),$$
$$G_N = J_\infty^{-1}(I_\infty(M_1)),$$

and

$$I_\infty(G_M) = J_\infty(G_N) = I_\infty(M_1) \cap J_\infty(N_1).$$

In particular, G_M and G_N are Borel sets.

Lemma 10.4.2. *We have*

$$\mu(M_1 \setminus G_M) = \nu(N_1 \setminus G_N) = 0.$$

Proof. Let K_1 be an arbitrary compact set in N_1. Suppose that there exists a compact set $F \subset M_1$ such that $\mu(F) > 0$ and $F \cap G_M$ is an empty set. Note that $J_n^{-1}(I_n(F))$ contains $J_{n+1}^{-1}(I_{n+1}(F))$. Set

$$A_n = \nu(J_n^{-1}(I_n(F)) \cap K_1).$$

From the assumptions on the set F, we find that the set

$$\bigcap_{n=1}^{\infty} J_n^{-1}(I_n(F))$$

is an empty set. Therefore we have $A_n \to 0$ as $n \to \infty$. Now

$$\mu(F) \leq \mu(I_n^{-1}(I_n(F))) = \int_{I_n^{-1}(I_n(F))} \mathrm{d}\mu = \int_{J_n^{-1}(I_n(F))} \mathrm{d}\nu = \nu(J_n^{-1}(I_n(F))).$$

The lim inf of the sequence of values of the integrals on the right-hand side, as $n \to \infty$, is at most equal to $\nu(N_1 \setminus K_1)$. By varying the set K_1, we obtain a contradiction. Therefore, $u(M_1 \setminus G_M) = 0$. Similarly, $\nu(N_1 \setminus G_N) = 0$. □

If $p \in G_M$, we have that

$$\bigcup_{n=1}^{\infty} J_n^{-1}(I_n(p))$$

is a non-empty subset of N_1. From Lemma 10.4.1 we see that there exists a unique point $q \in N_1$ such that

$$\bigcup_{n=1}^{\infty} J_n^{-1}(I_n(p)) = \{q\}.$$

Set $\beta(p) = q$. We conclude that β is a well-defined bijection between G_M and G_N. We also have $\beta = J_\infty^{-1} \circ I_\infty$.

Lemma 10.4.3. *The map $\beta : G_M \to G_N$ satisfies the following conditions.*

(a) *The map β is a bijection, and both β and β^{-1} are measurable maps. Also, β and β^{-1} map sets of measure zero onto sets of measure zero.*

(b) *The equality*

$$\frac{T(\psi)}{T(\phi)}(\beta) = \frac{\psi}{\phi}$$

holds on G_M for every $\phi, \psi \in A^1(M)$.

(c) *Let $p \in G_M$, and let $q \in N_1$. If*

$$\frac{T(\psi)}{T(\phi)}(q) = \frac{\psi}{\phi}(p)$$

holds for every $\phi, \psi \in A^1(M)$, then $q = \beta(p)$.

(d) *Let $q \in G_N$, and let $p \in M_1$. If*

$$\frac{T(\psi)}{T(\phi)}(q) = \frac{\psi}{\phi}(p)$$

holds for every $\phi, \psi \in A^1(M)$, then $p = \beta^{-1}(q)$.

(e) *If F is a measurable subset of G_M, then*

$$\int_F |\phi| = \int_{\beta(F)} |T(\phi)|$$

holds for every $\phi \in A^1(M)$. In particular, we have $\mu(F) = \nu(\beta(F))$.

Proof. It follows from Lemma 10.4.1 that β is a bijection between G_M and G_N. From the definition of G_M, G_N, I_n, J_n, and since ϕ_n is dense in $A^1(M)$, we see that

$$\frac{T(\phi)}{T(\phi_0)}(\beta) = \frac{\phi}{\phi_0}$$

on G_M. Since for $\phi, \psi \in A^1(M)$ we have

$$\frac{\phi/\phi_0}{\psi/\phi_0} = \frac{\phi}{\psi},$$

it follows that β satisfies condition (b). The fact that β satisfies conditions (c) and (d) follows directly from Lemma 10.4.1.

We have to show that β satisfies condition (e). Let F be a compact set in G_M. Let $u_n : \mathbb{C}^n \to \mathbb{R}$ be the function given by

$$u_n(w_1, \ldots, w_n) = |w_1|,$$

where $(w_1, \ldots, w_n) \in \mathbb{C}^n$. Let v_n be the function defined as the product of u_n and the characteristic function of the set $I_n(F)$. We see that

$$\int_{I_n^{-1}(I_n(F))} |\phi_1| \, dx \, dy = \int_{J_n^{-1}(I_n(F))} |T(\phi_1)| \, dx_1 \, dy_1.$$

In particular, since $\beta(F)$ is a Borel set, we have that the above integrals are well defined. Since $I_{n+1}^{-1}(I_{n+1}(F))$ is contained in $I_n^{-1}(I_n(F))$, from Lemma 10.4.1, we have that the set

$$\bigcup_{n=1}^{\infty} I_n^{-1}(I_n(F))$$

is equal to F. Now, from Lemma 10.4.2 we conclude that

$$\left| \int_{I_n^{-1}(I_n(F))} |\phi_1| \, dx \, dy - \int_F |\phi_1| \, dx \, dy \right| \to 0$$

as $n \to \infty$. Similarly,

$$\left| \int_{J_n^{-1}(I_n(F))} |T(\phi_1)| \, dx \, dy - \int_{\beta(F)} |T(\phi_1)| \, dx_1 \, dy_1 \right| \to 0$$

as $n \to \infty$. So far, we have obtained (e) whenever F is a compact subset of G_M. Applying the standard argument, we obtain (e) for every measurable set F contained in G_M. From (e) we also conclude that β maps the sets of measure zero onto sets of measure zero. This proves (a).

In the same way, we can show that (e) holds for every ϕ_n, and from the density of the sequence $\{\phi_n\}$, we have that (e) holds for every $\phi \in A^1(M)$. If we take $\phi = \phi_0$, we see that β and β^{-1} are measure-preserving maps, that is, $\mu(F) = \nu(\beta(F))$ for every measurable set F in G_M. □

We now have to show that there exists a conformal map $\alpha : M \to N$ such that $\alpha = \beta$ almost everywhere on M.

Definition 10.4.2. *We say that a point $p \in M$ belongs to the set E_M if there exists an open set U containing p and a holomorphic injection $\alpha : U \to N$, such that the equality*

$$\frac{T(\psi)}{T(\phi)} \circ \alpha = \frac{\psi}{\phi} \tag{10.15}$$

holds on U for every $\phi, \psi \in A^1(M)$. In a similar way, we define the set E_N.

It follows that the set E_M is an open subset of M and E_N is an open subset of N. We will show that the map α is a well-defined holomorphic bijection between E_M and E_N.

To show that α is well defined, we need to consider two different open sets U_1 and U_2 and two holomorphic injections $\alpha_1 : U_1 \to N$ and $\alpha_2 : U_2 \to N$, which both satisfy (10.15), and to show that α_1 and α_2 agree on $U_1 \cap U_2$. Combining the fact that $M \setminus M_1$ is a discrete set, and (c) and (d) from Lemma 10.4.3, together with (10.15), we conclude that $\alpha_k = \beta$ almost everywhere on U_k for $k = 1, 2$. This implies that α_1 and α_2 agree almost everywhere on $U_1 \cap U_2$. Similarly, we show that α is a bijection between E_M and E_N.

Lemma 10.4.4. *We have the following:*

(a) *Let $p_0 \in M_1$ and let $\{p_n\}$, $n \in \mathbb{N}$, be a sequence of points such that $p_n \in G_M$ and $p_n \to p_0$. If there is a point $q_0 \in N_1$ such that $\beta(p_n) \to q_0$, then the points p_0 and q_0 belong to E_M and E_N respectively.*

(b) *The sets E_M and E_N have full measure in M and N respectively. The map $\alpha : E_M \to E_N$ is a holomorphic bijection between E_M and E_N.*

(c) *We have that if $p \in E_M \cap M_1$ and $\alpha(p) \in N_1$, then $p \in G_M$. Similarly, if $q \in E_N \cap N_1$ and $\alpha^{-1}(q) \in M_1$, then $q \in G_N$. The maps α and β agree almost everywhere on M. Also α^{-1} and β^{-1} agree almost everywhere on N.*

(d) *The equality*

$$|(T(\phi) \circ \alpha)(\alpha')^2| = |\phi|$$

holds on E_M for every $\phi \in A^1(M)$.

Proof. Since $p_n \in G_M$, from property (b) of Lemma 10.4.3, we conclude that

$$\frac{T(\psi)}{T(\phi)}(\beta(p_n)) = \frac{\psi}{\phi}(p_n) \tag{10.16}$$

for every $\psi, \phi \in A^1(M)$. Since $p_n \to p_0 \in M_1$ and $\beta(p_n) \to q_0 \in N_1$, we have

$$\frac{T(\psi)}{T(\phi)}(q_0) = \frac{\psi}{\phi}(p_0)$$

for every $\phi, \psi \in A^1(M)$. On the other hand, from the definition of M_1 and N_1, we know that there are open neighbourhoods U_0 and V_0 of points p_0 and q_0, respectively, such that the functions $(\phi_1/\phi_0)(p)$ and $(T(\phi_1)/T(\phi_0))(q)$ are univalent as maps from U_0 and V_0, respectively, into $\mathbb{C}$. By modifying the neighbourhoods U_0 and V_0 if necessary, we see that there exists a conformal map $\widetilde{\alpha} : U_0 \to V_0$ such that the relation

$$\frac{T(\phi_1)}{T(\phi_0)} \circ \widetilde{\alpha} = \frac{\phi_1}{\phi_0} \tag{10.17}$$

holds on U_0. Here we choose U_0 and V_0 small enough so that they belong to M_1 and N_1 respectively.

Since $\widetilde{\alpha}$ is conformal on U_0, it follows from (10.16) and (10.17) that we have $\widetilde{\alpha}(p_n) = \beta(p_n)$ for all but finitely many $n \in \mathbb{N}$. Also, the equality

$$\frac{T(\psi)}{T(\phi)}(\widetilde{\alpha}(p_n)) = \frac{\psi}{\phi}(p_n)$$

holds for all but finitely many n and for every $\phi, \psi \in A^1(M)$. Since $p_n \to p_0$, we conclude that

$$\frac{T(\psi)}{T(\phi)}(\widetilde{\alpha}(p)) = \frac{\psi}{\phi}(p)$$

for every $p \in U_0$ and for every $\phi, \psi \in A^1(M)$. From property (c) of Lemma 10.4.3 and from the above equality, together with the fact that U_0 is a subset of M_1 and V_0 is a subset of N_1, we conclude that β is defined in U_0, and that U_0 is an open subset of G_M. This proves property (a).

The map β is a measurable map. In fact, if we ignore some sets of measure zero, we can think of β as a map from the interior of a fundamental polygon for M to the interior of such a polygon for N. This makes β an L^1-function on a plane region, and we can talk about the set of Lebesgue points of the map β on the interior of the fundamental polygon for M. The corresponding points on the surface M are called the Lebesgue points for β. Note that this definition of the Lebesgue points on M for β depends on the choice of the fundamental polygon for M. Nevertheless, this set has full measure in M.

Since β maps sets of positive measure onto sets of positive measure, we see that for almost every Lebesgue point $p_0 \in G(M)$ there exists $q_0 \in G_N$ and a sequence of distinct points $\{p_n\}$, $n \in \mathbb{N}$, such that $p_n \in G_M$, $p_n \to p_0$ and $\beta(p_n) \to q_0$. Therefore every such Lebesgue point is contained in E_M, and thus E_M has full measure in M.

We have already shown that α is the well-defined holomorphic bijection between E_M and E_N. Also, if $p \in E_M \cap M_1$ and $\alpha(p) \in N_1$, it follows from the definition of G_M that $p \in G_M$. Similarly, we have a corresponding result for $E_N \cap N_1$. We have seen that α agrees with β almost everywhere on any open subset of E_M. Since E_M has full measure in M, we conclude that α and β agree almost everywhere on M. In the same way, we see that α^{-1} agrees with β^{-1} almost everywhere on N.

For every $\phi \in A^1(M)$, we define

$$L(\phi) = (T(\phi) \circ \alpha)(\alpha')^2.$$

The differential $L(\phi)$ is defined on E_M. Also, it follows from (e) of Lemma 10.4.3 that for every measurable set F which is a subset of E_M, we have

$$\int_F |\phi| \,\mathrm{d}x\,\mathrm{d}y = \int_F |L(\phi)| \,\mathrm{d}x\,\mathrm{d}y. \tag{10.18}$$

Since ϕ and $L(\phi)$ are holomorphic forms on E_M, and E_M is an open set in M, it follows from (10.18) that

$$|(T(\phi) \circ \alpha)(\alpha')^2| = |\phi|$$

holds everywhere on E_M and for every $\phi \in A^1(M)$. □

Lemma 10.4.5. *Let $p' \in M$, and assume that there exists an open set U which contains p' such that $U \setminus \{p'\}$ is contained in E_M. Then $p' \in E_M$.*

Proof. Denote by $\widetilde{\alpha}$ the extension of the map α on the set U. Since $U \setminus \{p\}$ is a subset of E_M, we have that $\widetilde{\alpha}$ satisfies (10.15) on U as well, and by the definition of E_M, we have $p' \in E_M$. So, we may assume that $\widetilde{\alpha}(p')$ is not in N. Then $\widetilde{\alpha}(p')$ has to be a puncture in the boundary of N.

It is well know that there exists $\phi \in A^1(N)$ such that ϕ has a first-order pole at a given puncture in the boundary of M, in this case the puncture $\widetilde{\alpha}(p')$. From (d) in Lemma 10.4.4, we see that the equality

$$|(\phi \circ \widetilde{\alpha})(\widetilde{\alpha}')^2| = |T^{-1}(\phi)|$$

holds on $U \setminus \{p'\}$ and therefore on U as well. Since $T^{-1}(\phi)$ is holomorphic at p', we obtain a contradiction. □

Lemma 10.4.6. *The sets $M \setminus E_M$ and $N \setminus E_N$ have no isolated points. Furthermore, assume that (p_n), $n \in \mathbb{N}$, is a sequence of distinct points such that $p_n \in E_M$ for all n, and such that $p_n \to p_0 \in M_1 \setminus E_M$. Then the sequence $(\alpha(p_n))$ has no cluster points in N_1. The same conclusion holds for the set E_N.*

Proof. The fact that $M \setminus E_M$ and $N \setminus E_N$ have no isolated points follows from Lemma 10.4.5. Assume that the sequence $(\alpha(p_n))$ has a cluster point $q \in N_1$. By passing to a subsequence if necessary, we may assume $\alpha(p_n) \to q_0$, where $q_0 \in N_1$. Since $p_0 \in M_1$, from property (a) of Lemma 10.4.4 we see that $p_0 \in E_M$, which is a contradiction and completes the proof. □

We now consider the boundary behaviour of α. Let $\pi : \mathbb{D} \to N$ be the universal covering map, and let (z_n), $n \in \mathbb{N}$, be a sequence of distinct points in $\mathbb{D}$ such that

(a) $|z_n| \to 1$ as $n \to \infty$,

(b) there exists a sequence $r_n > 1$ such that $r_n \to \infty$ as $n \to \infty$, and the covering map $\pi : \mathbb{D} \to N$ is univalent in the hyperbolic disk of radius r_n, centred at z_n,

(c) if by q_n we denote the points on N such that $q_n = \pi(z_n)$, then there exists a sequence of distinct points (p_n), $p_n \in E_M \cap M_1$, for every $n \in \mathbb{N}$, such that $\alpha(p_n) = q_n$.

For every $p \in M$, there exists $\phi \in A^1(M)$ such that $\phi(p) \neq 0$. For each p_n as defined above, denote by $Z_n(M)$ the space of all $\phi \in A^1(M)$ such that $\phi(p_n) = 0$. Also for every q_n, we denote by $Z_n(N)$ the space corresponding to q_n. We see that $Z_n(M)$ is a vector subspace of $A^1(M)$ of codimension 1, and $Z_n(N)$ is a subspace of $A^1(N)$ of codimension 1.

We have $\alpha(p_n) = q_n$. Since $\phi_0 \neq 0$ on M_1, from (10.15), we conclude that $T(Z_n(M)) = Z_n(N)$. Let $\lambda_n : A^1(M) \to \mathbb{C}$ be a linear functional with norm equal to 1 which annihilates $Z_n(M)$. Let $\lambda'_n : A^1(N) \to \mathbb{C}$ be the linear functional such that $\lambda_n(\phi) = \lambda'_n(T(\phi))$ for every $\phi \in A^1(M)$. Since T is a linear isometry, λ'_n is of norm 1 and, by definition, λ'_n annihilates $Z_n(N)$.

We will show that from the assumption on z_n, it follows that for every $\phi \in A^1(N)$ we have $\lambda'_n(\phi) \to 0$.

From assumption (b) on z_n, it follows that the covering map π is univalent on the hyperbolic disk $\mathbb{D}_n$ centred at z_n of hyperbolic radius 1. Let B_n be a Möbius transformation of $\mathbb{D}$ onto itself such that $B_n(z_n) = 0$. Define the function η_n to be equal to $\overline{(B'_n)}/B'_n$ on $\mathbb{D}_n$. For $A \in \Gamma_N$, the covering group of $\mathbb{D}$ over N, set η_n to be equal to

$$\left(\frac{\overline{B'_n}}{B'_n} \circ A\right) \frac{\overline{A'_n}}{A'_n}$$

on the disk $A^{-1}(\mathbb{D}_n)$. Elsewhere in $\mathbb{D}$, we define η_n to be zero.

Denote by $\Theta : A^1(\mathbb{D}) \to A^1(N)$ the standard Poincaré operator. The operator Θ is surjective and has Banach norm at most 1. The covering map π is univalent on $\mathbb{D}_n$, and since η_n is equivariant, we see that η_n represents the lift of a certain $(-1, 1)$ measurable form on N.

Let $\omega \in A^1(\mathbb{D})$. From the mean value theorem for analytic functions, we obtain

$$\int_{\mathbb{D}} \eta_n \omega = C \sum_{A \in \Gamma_N} (\omega \circ A)(A')^2(z_n),$$

where $C > 0$ is some constant. Therefore, we have

$$\int_{\mathbb{D}} \eta_n \omega = 0$$

if and only if $\Theta(\omega) \in Z_n(N)$ for $\omega \in A^1(\mathbb{D})$. Therefore there exists $t_n \in \mathbb{C}$ such that for each $\omega \in A^1(\mathbb{D})$ we have

$$\int_{\mathbb{D}} t_n \eta_n \omega = \lambda_n'(\Theta(\omega)). \tag{10.19}$$

Set $\omega_n = (B_n')^2$. We have that the Banach norm $||\omega_n||_1 = \pi$. Also, from the properties of z_n, it follows that

$$\int_{\mathbb{D}} \eta_n \omega_n = \int_{\mathbb{D}_n} \eta_n \omega_n + \sum_{A\omega \in (\Gamma_N \setminus Id)} \int_{A(\mathbb{D}_n)} \eta_n \omega_n.$$

From the properties of z_n, it also follows that as $n \to \infty$ the hyperbolic distance between z_n and any point $A(z_n)$, for $A \in \Gamma_n \setminus Id$, converges uniformly to ∞. From the construction of ω_n, we have

$$\sum_{A\omega \in (\Gamma_N \setminus Id)} \int_{A(\mathbb{D}_n)} |\omega_n| \to 0,$$

and

$$\left| \int_{\mathbb{D}} \eta_n \omega_n \right| \to C_1$$

as $n \to \infty$ and where C_1 stands for the Euclidean area of the hyperbolic disk of hyperbolic radius 1, centred at 0.

We have constructed the sequence $\varphi_n = \Theta(\omega_n) \in A^1(N)$ so that $||\varphi_n||_1 \le \pi$ and

$$\lim_{n\to\infty} \frac{|\lambda'(\varphi_n)|}{|t_n|} = C_1.$$

Since $|\lambda_n'(\varphi_n)| \le \pi$, we conclude that $\limsup |t_n| \le \frac{\pi}{C_1}$.

We have that $t_n \eta_n$ is uniformly bounded on $\mathbb{D}$ (the bound is independent of n), and we have that $t_n \eta_n$ converges to zero pointwise on $\mathbb{D}$. This conclusion, together with (10.19), yields that for any fixed $\omega \in A^1(\mathbb{D})$, we have

$$\int_{\mathbb{D}} t_n \eta_n \omega = \lambda_n'(\Theta(\omega)) \to 0$$

as $n \to \infty$. Since Θ is surjective, $\lambda_n'(\phi) \to 0$ as $n \to \infty$ for every $\phi \in A^1(N)$.

Lemma 10.4.7. *With the notation and assumptions as above, the sequence p_n has no cluster points in M_1.*

Proof. Assume that p_n does have a cluster point in M_1. By passing to a subsequence if necessary, we may assume that there exists a point $p_0 \in M_1$ such that $p_n \to p_0$. Let $\psi_0 \in A^1(M)$ such that $\psi_0(p_0) \ne 0$. We will show that the sequence $\lambda_n(\psi_0)$ does not converge to zero.

To see that $\lambda_n(\psi_0)$ does not converge to zero, note that for large enough n, the functional λ_n is a multiple of the linear functional

$$\psi \mapsto \frac{\psi}{\psi_0}(p_n),$$

for $\psi \in A^1(M)$. Since this functional converges to the linear functional

$$\psi \mapsto \frac{\psi}{\psi_0}(p_0),$$

for $\psi \in A^1(M)$, which is a non-zero linear functional, we see that $\lambda_n(\psi_0)$ does not converge to zero. □

Now let $p_0 \in E_M$, and let $E_{M,0}$ be the maximal connected component of the set E_M which contains p_0.

Lemma 10.4.8. *Let γ be a Jordan curve in $E_{M,0}$ such that the Jordan domain Ω_0 bounded by γ is a well-defined planar subregion of M_1, and such that $p_0 \in \Omega_0$. Then Ω_0 is contained in $E_{M,0}$.*

The techincal proof of this lemma can be found in [35] and is omitted. Let $\mathbb{D}_0$ be a hyperbolic disk centred at p_0 which is contained in a larger concentric hyperbolic disk which is injectively embedded in M_1. Set $\Omega = \mathbb{D}_0 \cap E_{M,0}$, and denote by Ω_0 the connected component of the set Ω which contains p_0. We have that Ω_0 is an open connected subset of $\mathbb{D}_0$.

Lemma 10.4.9. *We have $\mathbb{D}_0 = \Omega_0$.*

Proof. We assume that $\mathbb{D}_0 \neq \Omega_0$. It follows from Lemma 10.4.8 that Ω_0 is a simply connected region. Set $C = E_M \cap \partial\Omega_0$ and set $B = \partial\Omega_0 \setminus C$. From the construction, we have that C is an open subset of the circle $\partial\mathbb{D}_0$. Thus B is a compact set in M.

Suppose that a point $q \in N$ belongs to the boundary of $\alpha(\Omega_0)$. If, in addition, we assume that q does not belong to E_N, then from Lemma 10.4.6 we conclude that $q \in N \setminus N_1$. Therefore we have shown that if (p_n), $n \in \mathbb{N}$, is a sequence of points in Ω_0 such that $p_n \to B$, then the sequence $\alpha(p_n)$ has no cluster points in N_1. In particular, the only possible accumulation points (in N) of the sequence $\alpha(p_n)$ belong to the set $N \setminus N_1$.

Let $f : \mathbb{D} \to \Omega_0$ be a Riemann map with $f(0) = p_0$. Let C_1 be the open subset of the unit circle $\partial\mathbb{D}$ such that the map f extends analytically onto C_1, and let $f : C_1 \to C$. Denote by B_1 the complement of the set C_1 with respect to $\partial\mathbb{D}$.

First, consider the case when the 1-dimensional Lebesgue measure of the set B_1 is zero. We may assume that $\mathbb{D}_0$ is in the complex plane (by taking a lift of $\mathbb{D}_0$ in the universal cover of M). Since the measure of B_1 is zero, we have that C_1 is dense in $\partial\mathbb{D}$ and B_1 is a totally disconnected subset of $\partial\mathbb{D}$. But then the Riemann map $f : \mathbb{D} \to \Omega_0$, $f(0) = p_0$ can be conformally extended (by the reflection principle) onto $\mathbb{C} \setminus B_1$. Now the extended map $f : \mathbb{C} \setminus B_1 \to \mathbb{C}$ maps the region $\mathbb{C} \setminus B_1$ into the region $\mathbb{C} \setminus B$. Since B_1 is of measure zero, it follows that the extended map f is a Möbius transformation and that $f(\mathbb{D}) = \mathbb{D}_0$, which is a contradiction.

Next, we obtain a contradiction to the assumption that B_1 has positive measure. Set $\widetilde{g} = \alpha \circ f$. Since α maps the domain Ω_0 into N_1, we have that the map $\widetilde{g} : \mathbb{D} \to N_1$ is a holomorphic injection. Denote by $g_1 : \mathbb{D} \to \mathbb{D}$ the lift of the map $\widetilde{g}$ into the universal cover of the surface N_1 (note that the points in $N \setminus N_1$ are isolated points in the boundary of N_1). Let (z_n), $n \in \mathbb{N}$, be any sequence in $\mathbb{D}$ such that z_n converges to the set B_1. Then, as we have already pointed out, from Lemma 10.4.8 it follows that $\alpha(f(z_n))$ does not have cluster points in N_1. This implies that $|g_1(z_n)| \to 1$ as $n \to \infty$. This last fact, together with the fact that the measure of B_1 is positive, implies (see Corollary 6.20 of [42]) that the map g_1 is conformal at almost every point of the set B_1 (we again refer to [42] for the notion of being conformal at boundary points). Let X_1 be the subset of B_1 whose elements are the points where the map g_1 is conformal. We know that X_1 has full measure in B_1.

Let $x \in X_1$. From Proposition 4.10 of [42], we have that smooth curves in a Stolz angle at x are mapped onto smooth curves, and the angles between curves are preserved. In particular, for every sequence $z_n \in \mathbb{D}$ such that $z_n \to x$ and z_n is contained in a given Stolz angle at x, we have $r_n \to \infty$, where r_n is the radius of the maximal hyperbolic disk centred at $g_1(z_n)$ which is contained in $g_1(\mathbb{D})$. Let $\pi_1 : \mathbb{D} \to N_1$ be the corresponding universal covering map. Then, we conclude that the sequence $\pi_1(g_1(z_n))$ cannot accumulate at any isolated puncture in the boundary of N_1 (this includes the points of the set $N \setminus N_1$). This follows from the fact that if q_n is a sequence on a given Riemann surface which converges to a puncture in the boundary of that surface, then the injectivity radius at q_n goes to zero.

Denote by $g : \mathbb{D} \to \mathbb{D}$ the lift of the map $\widetilde{g}$ into the universal cover of the surface N, and let $\pi : \mathbb{D} \to N$ be the corresponding universal covering map (note here that the points in $N \setminus N_1$ are not in the boundary of N). Let $x \in X_1$, and let z_n be a sequence in $\mathbb{D}$, $z_n \to x$ and z_n is contained in a given Stolz angle at x. Since we have that the corresponding sequence $\pi(g(z_n)) = \pi_1(g_1(z_n))$ does not accumulate at any point of the set $N \setminus N_1$, and by using Lemma 10.4.8 again, we find that $|g(z_n)| \to 1$. This means that if the function g has a radial limit at some point $x \in X_1$, then this limit lies on $\partial\mathbb{D}$.

On the other hand, the map g has radial limits almost everywhere on $\partial\mathbb{D}$. Since X_1 is of positive measure and all the radial limits at points from X_1 lie on $\partial\mathbb{D}$, from Corollary 6.20 of [42], we conclude that the map g is conformal almost everywhere on X_1. Denote by X the subset of X_1 which consists of points where g is conformal. We know that X has full measure in X_1.

Now, again applying Proposition 4.10 of [42], we have that for every $x \in X$ and every sequence $z_n \in \mathbb{D}$, $z_n \to x$, where z_n is contained in a given Stolz angle at x, we have $r_n \to \infty$, where r_n is the radius of the maximal hyperbolic disk which is centred at $g(z_n)$ and contained in $g(\mathbb{D})$. However, the sequence $\alpha^{-1}(\pi(g(z_n)))$ remains in the disk $\mathbb{D}_0$, which is compactly contained in M_1. From Lemma 10.4.7 we obtain a contradiction.

We have therefore proved that $\Omega_0 = \mathbb{D}_0$. $\square$

Lemma 10.4.9 implies that every injectively embedded hyperbolic disk in M_1 belongs to E_M. Therefore we have that M_1 is contained in E_M. By Lemma 10.4.6, there are no isolated points in E_M, and therefore $E_M = M$ and, similarly, $E_N = N$. Therefore $\alpha : M \to N$ is a conformal map that by Lemma 10.4.4 satisfies the equality

$$|(T(\phi) \circ \alpha)(\alpha')^2| = |\phi|$$

for all $\phi \in A^1(M)$. Therefore, for every $\phi \in A^1(M)$, there exists a complex number θ with $|\theta| = 1$ such that the equality

$$(T(\phi) \circ \alpha)(\alpha')^2 = \theta\phi$$

holds on M. Since T is a linear map, we find that the corresponding θ is the same for all $\phi \in A^1(M)$. This proves the following generalization of Theorem 10.3.1.

Theorem 10.4.2. *Suppose that the Riemann surfaces M and N are not of exceptional type. Let $T : A^1(M) \to A^1(N)$ be a surjective linear isometry. Then the isometry T is geometric. The surface M and N are conformally related and therefore homeomorphic.*

10.5 Biholomorphic maps between Teichmüller spaces

In this chapter, we show that all biholomorphic maps between Teichmüller spaces are induced from a quasiconformal map between the Riemann surfaces in question. We begin with some definitions.

Definition 10.5.1. *The automorphism group* Aut($T(M)$) *of Teichmüller space $T(M)$ is the group of all biholomorphic maps from $T(M)$ onto itself.*

Definition 10.5.2. *Every quasiconformal map $g : M \to N$ between Riemann surfaces M and N induces a mapping $\rho_g : T(M) \to T(N)$ given by*

$$[\rho_g(f)] = [f \circ g^{-1}].$$

Such a mapping between Teichmüller spaces is called geometric.

Exercise 10.5.1. Show that a geometric mapping is biholomorphic and an isometry in the Teichmüller metric.

Definition 10.5.3. *The mapping class group* MC(M) *is the group of all classes of quasiconformal maps from the Riemann surface M onto itself. Further, every $g \in \mathrm{MC}(M)$ induces an automorphism ρ_g of $T(M)$.*

Definition 10.5.4. *Let M be a Rieman surface of non-exceptional type. We say that M has the isometry property if the following holds: if M_1, M_2 are any two Riemann surfaces which are quasiconformally related to M, then every surjective linear isometry $T : A^1(M_1) \to A^1(M_2)$ is geometric (note that we use the definition of geometric given in the previous section).*

We now state a proposition which shows why the notion of having the isometry property is important. The proof may be found in [14].

Proposition 10.5.1. *Let M and N be quasiconformally related Riemann surfaces which are not of exceptional type. If M has the isometry property, then every biholomorphic map between $T(M)$ and $T(N)$ is geometric.*

Theorem 10.5.1. *If M is a Riemann surface of non-exceptional type, then*

$$\mathrm{Aut}(T(M)) = \mathrm{MC}(M).$$

Proof. Theorem 10.3.1 says that every surjective linear isometry between Bergman spaces of Riemann surfaces is geometric. Therefore the proof follows by applying this and Proposition 10.5.1. □

Proposition 10.5.2. *Let M and N be Riemann surfaces of non-exceptional type. If $F : T(M) \to T(N)$ is biholomorphic, then there exists Riemann surfaces $\widetilde{M}$ and $\widetilde{N}$ which are quasiconformally related to M and N respectively, and there exists a linear isometry $T : A^1(\widetilde{M}) \to A^1(\widetilde{N})$.*

The proof of this proposition may be found in [14]. We can use this proposition to show that every biholomorphic map between Teichmüller spaces is geometric.

Theorem 10.5.2. *Let M and N be Riemann surfaces of non-exceptional type, and let $F : T(M) \to T(N)$ be biholomorphic. Then M and N are quasiconformally related and F is geometric.*

Proof. Applying Proposition 10.5.2 and Theorem 10.3.1, we find that

$$T : A^1(\widetilde{M}) \to A^1(\widetilde{N})$$

is geometric. Therefore M and N are conformally related, and in particular quasiconformally related. Now by Proposition 10.5.1, F is geometric, which completes the proof. □

11

Local rigidity of Teichmüller spaces

In the previous chapter, we saw that a surjective linear isometry between the Bergman spaces of two Riemann surfaces implies that the two Riemann surfaces are conformally equivalent. In this chapter, we will show that when two Riemann surfaces M and N have infinite dimensional Teichmüller spaces, their Bergman spaces will always be isomorphic. This then implies, via the Bers embedding, that their Teichmüller spaces will be locally bi-Lipschitz equivalent. We follow the presentation in [20].

11.1 Bergman kernels

Lemma 11.1.1. *The Bergman kernel on $\mathbb{D} \times \mathbb{D}$ is given by*

$$K(z, \zeta) = \frac{1}{(1 - z\bar{\zeta})^4},$$

and satisfies the following properties:

(a) $K(z, \zeta) = \overline{K(\zeta, z)}$,

(b) *for every Möbius transformation* $f : \mathbb{D} \to \mathbb{D}$,

$$K(f(z), f(\zeta)) f'(z)^2 \overline{f'(\zeta)^2} = K(z, \zeta),$$

(c) $\int_{\mathbb{D}} |K(z, \zeta)| \, \mathrm{d}x \, \mathrm{d}y \le \pi \rho^2(\zeta)$,

(d) *for every* $f \in A^1(\mathbb{D})$,

$$f(z) = \frac{3}{\pi} \int_{\mathbb{D}} \rho^{-2}(\zeta) K(z, \zeta) f(\zeta) \, \mathrm{d}\xi \, \mathrm{d}\eta$$

where $\rho(z) = (1 - |z|^2)^{-1}$ is the hyperbolic density on $\mathbb{D}$, and ξ, η are coordinates in the ζ-plane.

(e) *for each* $\zeta \in \mathbb{D}$,

$$\sup_{z \in \mathbb{D}} |K(z,\zeta)|\rho^{-2}(z) < \infty.$$

Proof. The first property is obvious from the definition of K. The second property follows from an elementary calculation. For the third property, consider

$$g(\zeta) = \int_{\mathbb{D}} |K(z,\zeta)| \, \mathrm{d}x \, \mathrm{d}y$$

and observe that under the change of variable $\zeta \mapsto f(\zeta)$, for a Möbius transformation $f : \mathbb{D} \to \mathbb{D}$, we have $g(f(\zeta))|f'(\zeta)|^2 = g(\zeta)$. Therefore $g(\zeta)$ can be determined by evaluating $g(0)$. Since $g(0) = \pi$, this gives the third property with equality. For the fourth property, consider first the mean value property for harmonic functions, that is for $r < 1$,

$$f(0) = \frac{1}{2\pi} \int_0^{2\pi} f(re^{i\theta}) \, \mathrm{d}\theta.$$

Hence

$$f(0) \int_0^1 (1-r^2) r \, \mathrm{d}r = \frac{1}{2\pi} \int_0^{2\pi} \int_0^1 (1-r^2) r f(re^{i\theta}) \, \mathrm{d}r \, \mathrm{d}\theta,$$

since f is integrable in $\mathbb{D}$. This can now be rewritten as

$$f(0) = \frac{3}{\pi} \int_{\mathbb{D}} \rho^{-2}(\zeta) K(0,\zeta) f(\zeta) \, \mathrm{d}\xi \, \mathrm{d}\eta,$$

and the invariance properties of ρ and K under Möbius transformations give the general formula. The final property is obvious from the definitions of K and ρ. □

Every hyperbolic Riemann surface M has the disk $\mathbb{D}$ as its universal cover, that is, there is a Fuchsian covering group G such that $M \simeq \mathbb{D}/G$. Now, given such a covering group G, form the Poincare theta series given by

$$F(z,\zeta) = \sum_{\gamma \in G} K(\gamma(z),\zeta)\gamma'(z)^2.$$

Lemma 11.1.2. *The series for* $F(z,\zeta)$ *converges absolutely and uniformly on compact subsets of* $\mathbb{D}$ *to a function holomorphic in* z, *antiholomorphic in* ζ, *and satisfies*

(a) $F(z,\zeta) = \overline{F(\zeta,z)}$,

(b) *for* $\gamma \in G$, $F(\gamma(z),\zeta)\gamma'(z)^2 = F(z,\zeta)$,

(c) *for* A *in the normalizer of* G, $F(A(z),A(\zeta))A'(z)^2\overline{A'(\zeta)^2} = F(z,\zeta)$,

(d) *for every holomorphic quadratic differential* $\psi \in A^1(\mathbb{D})$ *which is invariant under the group* G,

$$\psi(z) = \frac{3}{\pi} \int_{\Omega} \rho^{-2}(\zeta) F(z,\zeta)\psi(\zeta) \, \mathrm{d}\xi \, \mathrm{d}\eta,$$

where $\Omega \subseteq \mathbb{D}$ *is a fundamental region of* G,

(e) *for a fixed* $|\zeta| < 1$,

$$\sup_{z\in\mathbb{D}} |F(z,\zeta)\rho^{-2}(z)| < \infty.$$

Proof. The fact that $F(z,\zeta)$ converges absolutely and uniformly on compact subsets follows from the function $K(z,\zeta) = K_{\mathbb{D}}(z,\zeta)$ being integrable in z for fixed ζ and standard estimates. Also, F is clearly holomorphic in z and anti-holomorphic in ζ. Property (a) follows from the symmetry property of K and the fact that

$$K(\gamma(z),\gamma(\zeta))\gamma'(z)^2\overline{\gamma'(\zeta)^2} = K(z,\zeta).$$

Property (b) follows from the definition of (b). To prove property (c), let A be in the normalizer of G. Then

$$\begin{aligned} F(A(z),A(\zeta)) = A'(z)^2\overline{A'(\zeta)^2} &= \sum_{\gamma\in G} K(\gamma(A(z)),A(\zeta))\gamma'(A(z))^2A'(z)^2\overline{A'(\zeta)^2} \\ &= \sum_{\gamma\in G} K(\gamma(A(z)),A(\zeta))(\gamma\circ A)'(z)^2\overline{A'(\zeta)^2} \\ &= \sum_{\gamma\in G} K((A^{-1}\circ\gamma\circ A)(z),\zeta)(A^{-1}\circ\gamma\circ A)'(z)^2 \\ &= F(z,\zeta). \end{aligned}$$

The invariance property (b) of $F(z,\zeta)$ and the invariance of $\rho_{\mathbb{D}}$ means that the formula in (d) reduces to the case for $\mathbb{D}$, which was proved in Lemma 11.1.1. Finally, to prove (e), it is enough to show that

$$\sum_{\gamma\in G} |\gamma'(z)^2| \le C\rho_{\mathbb{D}}^2(z) \tag{11.1}$$

for some constant C depending only on the group G. Let

$$h(z) = \frac{1}{\pi}\int_{\omega} \frac{d\xi\, d\eta}{|1 - z\bar{\zeta}|^4},$$

where ω is a fundamental domain for G. Note that

$$\sum_{\gamma\in G} h(\gamma(z))|\gamma'(z)|^2 = \frac{1}{\pi}\int_{\mathbb{D}} \frac{d\xi\, d\eta}{|1 - z\bar{\zeta}|^4} = \frac{1}{(1-|z|^2)^2}.$$

The denominator of the integrand is less than $2^4 = 16$, and so

$$h(z) \ge \frac{1}{16\pi}|\omega|,$$

where $|\omega|$ is the Euclidean area of ω. This shows that (11.1) is true with

$$C = 16\pi(|\omega|)^{-1}.$$

□

For a hyperbolic Riemann surface $M \simeq \mathbb{D}/G$, there is a universal covering map $\pi : \mathbb{D} \to M$ such that $\pi \circ \gamma = \pi$, for all $\gamma \in G$. Pick a fundamental region ω of $\mathbb{D}/M$ so that $\pi|_\omega$ is injective, and denote now $\pi|_\omega$ by π without confusion. The hyperbolic density ρ_M for the surface M is defined by $\rho_M(\pi(z))|\pi'(z)| = \rho(z)$.

Definition 11.1.1. *The Bergman kernel function for M is given by*

$$K_M(\pi(z), \pi(\zeta))\pi'(z)^2\overline{\pi'(\zeta)^2} = F(z, \zeta).$$

Lemma 11.1.3. *The kernel function $K_M : M \times M \to \mathbb{C}$ defined above is holomorphic in the first argument, antiholomorphic in the second argument, and satisfies the following properties, where $\mu, \nu \in M$,*

(a) $K_M(\mu, \nu) = \overline{K_M(\nu, \mu)}$,

(b) *for every conformal* $f : M \to M$, $K_M(f(\mu), f(\nu))f'(\mu)^2\overline{f'(\nu)^2} = K_M(\mu, \nu)$,

(c) $\int_M |K_M(\mu, \nu)| \, d\mu \le \pi\rho_M^2(\nu)$,

(d) *for every* $\varphi \in A^1(M)$,

$$\varphi(\mu) = \frac{3}{\pi}\int_M \rho_M^{-2}(\nu)K_M(\mu, \nu)\varphi(\nu) \, d\nu,$$

(e) *for each fixed* $\nu \in M$,

$$\sup_{\mu \in M} |K_M(\mu, \nu)|\rho_M^{-2}(\mu) < \infty.$$

Proof. Most of the properties follow from the analogous properties of F, and here we will just prove the third property since it will be used shortly.

$$\int_M |K_M(\pi(z), \nu)| \, d\nu = \int_\omega |K_M(\pi(z), \pi(\zeta))||\pi'(\zeta)^2| \, d\zeta$$

where $\nu = \pi(\zeta)$ for $\zeta \in \omega$, a fundamental region for M in $\mathbb{D}$,

$$\begin{aligned} &= \int_\omega |F(z, \zeta)||\pi'(z)^{-2}| \, d\zeta \le |\pi'(z)^{-2}| \int_{\mathbb{D}} |K(z, \zeta)| \, d\zeta \\ &\le \pi|\pi'(z)^{-2}|\rho(z)^2 = \pi\rho_M(\pi(z))^2, \end{aligned}$$

which completes the proof. □

Define the linear map $P : L^1(M) \to A^1(M)$ by

$$(P(\varphi))(\mu) = \frac{3}{\pi}\int_M \rho_M^{-2}(\nu)K_M(\mu, \nu)\varphi(\nu) \, d\nu \tag{11.2}$$

for $\mu, \nu \in M$. For any $\varphi \in L^1(M)$, it is clear that the integral formula for $P(\varphi)$ means that $P(\varphi)$ will be holomorphic, so the image of P is indeed $A^1(M)$.

Theorem 11.1.1. *There exists a bounded linear projection $\theta : L^1(M) \to A^1(M)$, given by $\theta : \varphi \mapsto P(\varphi)$ for $\varphi \in L^1(M)$.*

Proof. The map θ is clearly linear, and bounded, since

$$||P(\varphi)|| = \int_M |P(\varphi(\mu))| \, d\mu = \frac{3}{\pi} \int_M \left| \int_M \rho_M^{-2}(\nu) K_M(\mu,\nu)\varphi(\nu) \, d\nu \right| d\mu$$

$$\leq \int_M \left(\int_M |K_M(\mu,\nu)| \, d\mu \right) \rho_M^{-2}(\nu)|\varphi(\nu)| \, d\nu$$

by Fubini's theorem, which we can apply by the fifth property in Lemma 11.1.3, and then using the third property of Lemma 11.1.3 gives

$$||P(\varphi)|| \leq 3 \int_M |\varphi(\nu)| \, d\nu$$

Hence, $||\theta|| \leq 3$. The integral reproducing formula given in (11.2) shows that $\theta|_{A^1(M)}$ is the identity, $\theta^2 = \theta$, and so θ is a projection. □

11.2 Operators on $A^1(M)$

Lemma 11.2.1. *Let K be a precompact subset of $\mathbb{D}$. Then the restriction operator $R_K : A^1(\mathbb{D}) \to A^1(K)$, given by $R(f) = f|_K$ for $f \in A^1(\mathbb{D})$, is compact.*

Holomorphic functions are defined on open sets and K need not be open. However we can define $A^1(K)$ by letting $f \in A^1(K)$ if there exists an open neighbourhood U of K in $\mathbb{D}$ such that f is holomorphic in U and

$$\int_U |f| < \infty.$$

Proof. The restriction operator R_K is compact if and only if for every bounded sequence $f_n \in A^1(\mathbb{D})$, the sequence $R_K(f_n)$ has a convergent subsequence.

For $z \in \mathbb{D}$, let $d(z, \partial\mathbb{D})$ be the shortest Euclidean distance from z to the boundary of $\mathbb{D}$, and similarly let $d(K, \partial\mathbb{D}) = \inf_{z \in K} d(z, \partial\mathbb{D})$. The Cauchy integral formula gives

$$|f(z_0)| \leq \frac{1}{\pi t^2} \int_0^{2\pi} \int_0^t |f(z_0 + re^{i\theta})| r \, dr \, d\theta$$

for $t < d(z_0, \partial\mathbb{D})$ and $f \in A^1(\mathbb{D})$. Therefore

$$|f(z_0)|(\pi t^2) \leq \int_{\mathbb{D}} |f|,$$

and this holds for $t < d(z_0, \partial\mathbb{D})$, so in particular, for any $z_0 \in K$,

$$|f(z_0)| \leq \frac{1}{\pi (d(K, \partial\mathbb{D}))^2} \int_{\mathbb{D}} |f|. \tag{11.3}$$

Now, let f_n be a bounded sequence in $A^1(\mathbb{D})$ and without loss of generality, $||f_n|| \leq 1$ for all n. We can find a precompact subset $\Omega \subset \mathbb{D}$ such that $K \subset \Omega$. Since $||f_n|| \leq 1$, (11.3) implies that $|f_n(z)| \leq C_\Omega$ for all n, for all $z \in \Omega$, and where

$$C_\Omega = (\pi(d(K, \partial\mathbb{D}))^2)^{-1}.$$

This shows that $R_\Omega(f_n)$ is uniformly bounded on Ω, which means that $R_\Omega(f_n)$ is a normal family. Hence there is a subsequence $R_\Omega(f_{n_k})$ which converges uniformly on compact subsets of Ω, and in particular, uniformly on K. Uniform convergence implies convergence in the L^1 norm, so there exists some function $g \in A^1(K)$ such that $R_K(f_{n_k}) \to g$. □

Lemma 11.2.2. *Let Ω be a simply connected precompact subset of a Riemann surface M. Then given $\epsilon > 0$, there exists a projection $P : L^1(\Omega) \to L^1(\Omega)$ such that $||P|| < 1$,*

$$||P(f) - f|| < \epsilon,$$

for all $f \in A^1(\Omega)$, and $P(L^1(\Omega))$ is isometric to $(l^1)_n$, where $(l^1)_n$ is the n-dimensional subspace of l^1 with all terms, except possibly the first n being equal to 0.

Proof. We can for simplicity assume that Ω is a bounded simply connected plane domain. Subdivide Ω into a finite number of subsets, $\Omega_1, \dots, \Omega_n$. For a given $f \in L^1(\Omega)$, define λ_i to be $\int_{\Omega_i} f$. We have

$$\sum_{i=1}^{n} |\lambda_i| = \sum_{i=1}^{n} \left| \int_{\Omega_i} f \right| \leq \int_\Omega |f| < \infty.$$

Define the map $P : L^1(\Omega) \to L^1(\Omega)$ by

$$P(f) = \sum_{i=1}^{n} \frac{\lambda_i}{m(\Omega_i)} \mathbf{1}_{\Omega_i},$$

where $\mathbf{1}_{\Omega_i}$ denotes the indicator function of Ω_i, and m is the usual two-dimensional Lebesgue measure of Ω_i. The map P is clearly linear and bounded ($||P|| \leq 1$ in fact), and also a projection, since $P^2 = P$.

We can define a map $\mu : P(L^1(\Omega)) \to (l^1)_n$ given by

$$\mu(P(f)) = (\lambda_1, \dots, \lambda_n, 0, \dots).$$

Now, $||\mu(P(f))||_{l^1} = \sum_{i=1}^{n} |\lambda_i|$. Also,

$$||P(f)||_1 = \int_\Omega |P(f)| = \int_\Omega \left| \sum_{i=1}^{n} \frac{\lambda_i}{m(\Omega_i)} \mathbf{1}_{\Omega_i} \right| = \sum_{i=1}^{n} \int_{\Omega_i} \left| \frac{\lambda_i}{m(\Omega_i)} \right| = \sum_{i=1}^{n} |\lambda_i|$$

since the supports of $\mathbf{1}_{\Omega_i}$ are disjoint. Hence μ is isometric, and so $P(L^1(\Omega))$ is isometric to $(l^1)_n$.

We now have to show that we can find a fine enough subdivision of Ω so that for the corresponding projection P, $||P(f) - f|| < \epsilon$ for $f \in A^1(\Omega)$ with $||f|| \leq 1$. Since Ω is precompact in M,

$$\sup\{|f(z)|\}$$

is bounded, where the supremum is taken over all $f \in A^1(M)$ with $||f|| \leq 1$ and over all $z \in \Omega$ (recall the proof of Lemma 11.2.1). This means that

$$\Theta = \{f|_\Omega : f \in A^1(M), ||f|| \leq 1\}$$

is a normal family, and hence is equicontinuous, i.e. for all $f \in \Theta$ and for all $\epsilon > 0$, there exists a $\delta > 0$ such that if $|z - z_0| < \delta$, for $z, z_0 \in \Omega$, then $|f(z) - f(z_0)| < \epsilon$.

If $B(z_i, \delta)$ is a ball centred at z_i of Euclidean radius δ, then for any holomorphic function f,

$$\frac{1}{m(B(z_i,\delta))} \int_{B(z_i,\delta)} f = f(z_i).$$

If now Ω is subdivided into $\Omega_1, \ldots, \Omega_n$, with each $\Omega_i \subset B(z_i, \delta)$ for some z_i, and P is the corresponding projection to this subdivision, then

$$\int_{\Omega_i} |f - P(f)| \leq \int_{B(z_i,\delta)} |f(z) - f(z_i)| < \epsilon\, m(B(z_0, \delta))$$

recalling that $m(B(z_i, \delta))$ is the area of $B(z_i, \delta)$, and noting that the last inequality follows from the equicontinuity of Θ. Hence

$$\int_\Omega |f - P(f)| < \epsilon\, m(\Omega),$$

and since we are assuming that $m(\Omega)$ is finite, and ϵ can be made as small as wished, then we have the desired conclusion that $||P - I||$ can be as small as desired for P corresponding to a suitably fine subdivision of Ω. $\square$

Finally in this section, we need a lemma concerning projections on Banach spaces.

Lemma 11.2.3. *Let Y be a complemented subspace of a Banach space X, and let $T : Y \to X$ be a linear operator satisfying*

$$||T - I|_Y|| < \epsilon.$$

Then if ϵ is sufficiently small, $T(Y)$ is closed and complemented.

Proof. Let $S = I|_Y - T$. Then $||S|| < \epsilon$. Let also $P_Y : X \to Y$ be a projection, which is guaranteed to exist since Y is a complemented subspace of X. Define $\widetilde{S} : X \to X$ by

$$\widetilde{S} = S \circ P_Y.$$

Then $\widetilde{S}$ is an extension of S and

$$||\widetilde{S}|| \leq ||P_Y||.||S||.$$

Thus if $\epsilon < ||P_Y||^{-1}$, we have $||\widetilde{S}|| < 1$. Now set

$$\widetilde{T} = I - \widetilde{S} : X \to X.$$

Then $\widetilde{T}$ is an extension of T, and $||\widetilde{S}|| < 1$ implies that $\widetilde{T}$ is invertible and therefore a homeomorphism. We have $T(Y) = \widetilde{T}(Y)$ is closed and since $\widetilde{T}(\ker(P_Y))$ is complementary to $\widetilde{T}(\text{Im}(P_Y))$, $T(Y)$ is complemented. □

11.3 An isomorphism between $A^1(M)$ and l^1

Theorem 11.3.1. *If M is a hyperbolic Riemann surface of infinite analytic type, then $A^1(M)$ is isomorphic to the sequence space l^1.*

Proof. From Theorem 7.7.7, we know that this theorem applies to all Riemann surfaces where the dimension of $A^1(M)$ is infinite, for example, the plane punctured at the integer lattice points, or an infinite genus surface.

First of all, we subdivide M in an appropriate way. For every $p \in M$, there exists an open subset $U_p \subset M$ containing p, and a chart π_p such that $\pi_p(U_p)$ is a disk in $\mathbb{C}$ and $\pi_p(p) = 0$. Let V_p be an open simply connected set in M whose closure is contained in U_p, so that in particular $\pi_p(V_p)$ is a precompact subset of $\pi_p(U_p)$.

As p varies through M, $(V_p)_{p \in M}$ forms an open cover of M, and it is possible to find a countable subset $p_1, p_2, \ldots$, such that

$$M = \bigcup_{i=1}^{\infty} V_{p_i}$$

Now modify the subsets V_{p_i} to give a disjoint partition of M in the following way:

Define $M_1 = V_{p_1}$, and then inductively,

$$M_n = V_{p_n} \setminus \left(\bigcup_{i=1}^{n-1} V_{p_i} \right).$$

Let $R_i : L^1(M) \to L^1(M_i)$ be the restriction map given by $R_i(f) = f|_{M_i}$, for $f \in L^1(M)$.

Define the operator $R : L^1(M) \to (L^1(M_1) \oplus L^1(M_2) \oplus \cdots)_1$ by

$$R(f) = (R_1(f), R_2(f), \ldots),$$

for $f \in L^1(M)$. The operator R is isometric, since

$$||R(f)|| := \sum_{i=1}^{\infty} ||R_i(f)|| = \sum_{i=1}^{\infty} \int_{M_i} |f| = \int_M |f| = ||f||,$$

and R is also clearly surjective. Now, given $\epsilon_i > 0$, by Lemma 11.2.2, we can find a projection P_i of $L^1(M_i)$ into itself satisfying $||P_i|| \leq 1$, $P_i(L^1(M_i))$ is isometric to $(l^1)_{\alpha_i}$ for some $\alpha_i \in \mathbb{Z}^+$, and $||P_i(R_i(f)) - R_i(f)|| \leq \epsilon_i$ for all $f \in A^1(M)$ with $||f|| < 1$. Let

$$\Lambda = (P_1(L^1(M_1)) \oplus P_2(L^1(M_2)) \oplus \cdots)_1,$$

a subspace of $(L^1(M_1) \oplus L^1(M_2) \oplus \cdots)_1$. Since each $P_i(L^1(M_i))$ is isometric to $(l^1)_{\alpha_i}$ for some $\alpha_i \in \mathbb{Z}^+$, Λ is isometric to l^1. Now we define the operator $T : R(A^1(M)) \to \Lambda$ by

$$T(R_1(f), R_2(f), \ldots) = (P_1(R_1(f)), P_2(R_2(f)), \ldots).$$

Since the dimension of $A^1(M)$ is infinite, $R(A^1(M))$ also must be infinite dimensional. We also have

$$||T(\xi) - \xi|| \leq \left(\sum_{i=1}^{\infty} \epsilon_i\right) ||\xi||$$

for $\xi \in R(A^1(M))$, and so given $\epsilon > 0$, it is possible to choose the $(\epsilon_i)_i$ so that $||T(\xi) - \xi|| < \epsilon||\xi||$, for $\xi \in R(A^1(M))$.

There exists a bounded linear projection $\theta : L^1(M) \to A^1(M)$ by Theorem 11.1.1. Therefore, there is a bounded linear projection $\widetilde{\theta} : R(L^1(M)) \to R(A^1(M))$, given by

$$\widetilde{\theta}(R_1(f), R_2(f), \ldots) = (R_1(\theta(f)), R_2(\theta(f)), \ldots)$$

which is clearly linear, bounded and satisfies $\widetilde{\theta}^2 = \widetilde{\theta}$. Therefore $R(A^1(M))$ is complemented in $R(L^1(M))$. Thus, by Lemma 11.2.3, if ϵ is small enough, $T(R(A^1(M)))$ is complemented in $R(L^1(M))$ and, in particular, Λ. This follows since if $W \subset Y$ is complemented in X, then there exists a projection $S : X \to W$, $(\mathrm{Im}(S)) \cap Y$ is complemented in Y and so W is complemented in Y.

If $\epsilon < 1$, then $||T - I|| < 1$, and Lemma 11.2.3 gives that T is thus invertible and an isomorphism. By a classical result due to Pelczynski [41], every infinite dimensional complemented subspace of l^1 is isomorphic to l^1, and so $A^1(M)$ is isomorphic to l^1. □

Let $\alpha_M : A^1(M) \to l^1$ be the isomorphism in Theorem 11.3.1. This induces an isomorphism of the Banach duals $\alpha_M^* : (l^1)^* \to (A^1(M))^*$. It is well known that the Banach dual of l^1 can be identified with l^∞. Furthermore, let M be a hyperbolic Riemann surface. Then to each bounded linear functional Φ on $A^1(M)$, there corresponds a unique $g \in Q(M)$ such that

$$\Phi(\varphi) = \int_M \rho_\Omega^{-2}(w)\varphi(w)\overline{g(w)}\, du\, dv$$

for all $\varphi \in A^1(M)$. Moreover, $\frac{1}{3}||g||_Q \leq ||\Phi|| \leq ||g||_Q$. This result can be found in [37] or [48]. Thus, if M is a hyperbolic Riemann surface, the Banach dual of $A^1(M)$ can be identified with $Q(M)$. This immediately gives us the following results.

Theorem 11.3.2. *If M is a hyperbolic Riemann surface of infinite analytic type, then $Q(M)$ is isomorphic to the sequence space l^∞.*

Corollary 11.3.1. *If M and N are two hyperbolic Riemann surfaces of infinite analytic type, then $A^1(M)$ and $A^1(N)$ are isomorphic, and $Q(M)$ and $Q(N)$ are isomorphic.*

Exercise 11.3.1. The following is an open problem. Theorem 11.3.1 shows that there exists a constant C_M depending on M such that

$$\frac{||\alpha_M(\varphi)||_{l^1}}{C_M} \leq ||\varphi||_1 \leq C_M||\alpha_M(\varphi)||_{l^1}, \tag{11.4}$$

for all $\varphi \in A^1(M)$. Is there a universal constant C such that (11.4) holds, with C replacing C_M, independently of M?

11.4 Local bi-Lipschitz equivalence of Teichmüller spaces

We have the following situation,

$$\lambda_M : T(M) \hookrightarrow Q(M), \qquad \alpha_M^* : Q(M) \to l^\infty$$

where the image of the Bers embedding λ_M is contained in $Q(M)$. Since λ_M is a locally bi-Lipschitz mapping, there exists a neighbourhood, X_M, of the identity class in $T(M)$ such that $\lambda_M|_{X_M}$ is bi-Lipschitz. Since α_M^* is an isomorphism, X_M is mapped onto a neighbourhood of the origin of l^∞ by $\alpha_M^* \circ \lambda_M$. If

$$Y_M = (\alpha_M^* \circ \lambda_M)(X_M),$$

then X_M and Y_M are bi-Lipschitz equivalent.

Lemma 11.4.1. *If M and N are two hyperbolic Riemann surfaces with infinite dimensional Bergman spaces, then a neighbourhood of the identity class in $T(M)$ is bi-Lipschitz equivalent to a neighbourhood of the identity class in $T(N)$.*

Proof. Consider the neighbourhoods of the identity class in the respective Teichmüller spaces given by X_M and X_N, and consider their images in l^∞ under the respective maps $\alpha_M^* \circ \lambda_M$ and $\alpha_N^* \circ \lambda_N$, given by Y_M and Y_N.

$$T(M) \xrightarrow{\lambda_M} Q(M) \xrightarrow{\alpha_M^*} l^\infty \xleftarrow{\alpha_N^*} Q(N) \xleftarrow{\lambda_N} T(N)$$

The sets Y_M and Y_N are both open neighbourhoods of the origin in l^∞, and so $Y := Y_M \cap Y_N$ is also an open neighbourhood of the origin. Since $\alpha_M^* \circ \lambda_M$ is a bi-Lipschitz mapping of X_M, it has an inverse on Y, and

$$((\alpha_M^* \circ \lambda_M)^{-1})(Y) \subseteq X_M$$

is an open neighbourhood of the origin in $T(M)$.

Thus $(\alpha_N^* \circ \lambda_N) \circ (\alpha_M^* \circ \lambda_M)^{-1}$ is a bi-Lipschitz mapping from a neighbourhood of the identity class in $T(M)$, namely $((\alpha_M^* \circ \lambda_M)^{-1})(Y)$, to a neighbourhood of the identity class in $T(N)$, namely $(\alpha_N^* \circ \lambda_N)(Y)$. □

Theorem 11.4.1. *If M and N are two hyperbolic Riemann surfaces with infinite dimensional Bergman spaces, then their Teichmüller spaces are locally bi-Lipschitz equivalent.*

Proof. Let $M \simeq \mathbb{H}/G$, and identify $T(M)$ with $T(G)$. Recall that a chart for the neighbourhood of the identity class in $T(f_\mu \circ G \circ (f_\mu)^{-1})$ is a chart for the neighbourhood of $[\mu]$ in $T(M)$. Thus charts for any $[\mu] \in T(M)$ and $[\nu] \in T(N)$ correspond to charts for the respective identity classes in $T(f_\mu \circ \Gamma \circ (f_\mu)^{-1})$ and $T(f_\nu \circ \Gamma_1 \circ (f_\nu)^{-1})$.

Lemma 11.4.1 gives a bi-Lipschitz mapping between neighbourhoods of these two identity classes, and hence we have a bi-Lipschitz mapping between neighbourhoods of $[\mu] \in T(M)$ and $[\nu] \in T(N)$. □

References

[1] Abikoff, W., Earle, C., and Mitra, S. (2004) Barycentric extensions of monotone maps of the circle. *Contemp. Math.*, **355**, 1–20.

[2] Ahlfors, L. (1953/4) On quasiconformal mappings. *Journal d'Analyse Math.*, **3**, 1–58.

[3] Ahlfors, L. (1966) *Lectures on quasiconformal mappings.* Van Nostrand.

[4] Anderson, J. (1999) *Hyperbolic geometry.* Springer-Verlag.

[5] Astala, K. (1994) Area distortion of quasiconformal mappings. *Acta. Math.*, **173**, 37–60.

[6] Bers, L. and Royden, H. (1986) Holomorphic families of injections. *Acta. Math.*, **157**, 243–257.

[7] Beurling, A. and Ahlfors, L. (1956) The boundary correspondence under quasiconformal mappings. *Acta Math.*, **96**, 125–142.

[8] Bozin, V., Lakic, N., Markovic, V., and Mateljevic, M. (1998) Unique extremality. *Journal d'Analyse*, **75**, 299–338.

[9] Bozin, V., Markovic, V., and Mateljevic, M. (1998) Unique extremality in the tangent space of the universal Teichmüller space. *Integral Transforms and Special Functions*, **6**, 145–149.

[10] Carleson, L. (1967) On mappings, conformal at the boundary. *J. Analyse Math.* **19**, 1–13.

[11] Chirka, E. (2004) On the extension of holomorphic motions. *Doklady Math.*, **70**, no. 1, 516–519.

[12] Douady, A. and Earle, C. (1986) Conformally natural extension of homeomorphisms of the circle. *Acta. Math.*, **157**, no. 1–2, 23–48.

[13] Earle, C. (1969) On holomorphic cross-sections in Teichmüller spaces. *Duke Math. J.*, **36**, no. 2, 409–415.

[14] Earle, C. and Gardiner, F. (1996) Geometric isomorphism between infinite dimensional Teichmüller spaces. *Trans. Amer. Math. Soc.*, **348**, no. 3, 1163–1190.

[15] Earle, C., Kra, I., and Krushkal, S. (1994) Holomorphic motions and Teichmüller spaces. *Trans. AMS*, **343**, no. 2, 927–948.

[16] Earle, C. and McMullen, C. Quasiconformal Isotopies. *Holomorphic functions and moduli, Vol. I* (Berkley, CA, 1986), *Math. Sci. Res. Inst. Publ.*, **10**, 143–154, Springer, New York, 1988.

[17] Earle, C. and Markovic, V. (2003) Isometries between the spaces of L^1 holomorphic quadratic differentials on Riemann surfaces of finite type. *Duke Math. J.*, **120**, no. 2, 433–440.

[18] Epstein, D., Marden, A., and Markovic, V. (2004) Quasiconformal homeomorphisms and the convex hull boundary. *Annals of Mathematics*, **159**, no. 1, 305–336.

[19] Eremenko, A. and Hamilton, D. (1995) On the area distortion by quasiconformal mappings. *Proc. AMS*, **123**, no. 9, 2793–2797.

[20] Fletcher, A. (2006) Local rigidity of infinite dimensional Teichmüller spaces. *Journal LMS* (to appear).

[21] Gardiner, F. (1987) *Teichmüller theory and quadratic differentials.* John Wiley & Sons.

[22] Gardiner, F. and Lakic, N. (2000) *Quasiconformal Teichmüller theory. AMS.*

[23] Gardiner, F. and Sullivan, D. (1992) Symmetric structures on a closed curve, *Amer. J. Math.*, **114**, no. 4, 683–736.

[24] Gehring, F. and Reich, E. (1966) Area distortion under quasiconformal mappings. *Ann. Acad. Sci. Fenn. Ser. A I*, **388**, 15.

[25] Grötzsch, H. (1932) Übber möglichst konforme Abblidungen von schlichten Bereichen. *Ber. Verh. Sächs. Akad. Wiss. Leipzig*, **84**, 114–120.

[26] Hurwitz, A. (1893) Über algebraische Gebilde mit eindeutigen Transformationen in sich. *Math. Annalen*, **41**, 402–442.

[27] Imayoshi, Y. and Taniguchi, M. (1992) *An introduction to Teichmüller spaces.* Springer-Verlag.

[28] Kobayashi, S. (1970) *Hyperbolic manifolds and holomorphic mappings.* Marcel-Dekker, New York.

[29] Kra, I. (1972) *Automorphic forms and Kleinian groups.* Benjamin, Reading, Massachusetts.

[30] Lakic, N. (1997) Strebel points. *Contemporary Math.*, **211**, 417–431.

[31] Lehto, O. (1987) *Univalent functions and Teichmüller spaces.* Springer-Verlag.

[32] Lehto, O. and Virtanen, K. (1973) *Quasiconformal mappings in the plane.* Springer-Verlag.

[33] McMullen, C. (2004) *Riemann surfaces, dynamics and geometry. Course Notes* available at http://www.math.harvard.edu/ ctm/home/text/class/harvard/275/rs/rs.pdf

[34] Mane, R., Sad, P., and Sullivan, D. (1983) On dynamics of rational maps *Ann. Sci. Ecole. Norm. Sup.*, **16**, 193–217.

[35] Markovic, V. (2003) Biholomorphic maps between Teichmüller spaces. *Duke Math. J.*, **120**, no. 2, 405–431.

[36] Markovic, V. (2006) Quasisymmetric groups. *Journal Amer. Math. Soc.*, **19**, no. 3, 673–715.

[37] Mateljević, M. (1994) The dual of the Bergman space defined on a hyperbolic plane domain. *Publications de l'Institut Mathématique*, **56**, 135–139.

[38] Mostow, G. (1968) Quasi-conformal mappings in n-space and the rigidity of the hyperbolic space forms. *Publ. Math. IHES*, **34**, 53–104.

[39] Niebur, D. and Sheingorn, M. (1977) Characterization of Fuchsian groups whose integrable forms are bounded. *Ann. Math. (2)*, **106**, no. 2, 239–258.

[40] Pansu, P. (1989) Métriques de Carnot-Carathodory et quasiisométries des espaces symétriques de rang un. [Carnot-Caratheodory metrics and quasi-isometries of rank-one symmetric spaces]. *Ann. of Math. (2)*, **129**, no. 1, 1–60.

[41] Pelczynski, A. (1960) Projections in certain Banach spaces. *Studia. Math.*, **19**, 209–228.

[42] Pommerenke, C. (1992) *Boundary behaviour of conformal maps.* Springer-Verlag.

[43] Reich, E. (2002) Extremal Quasiconformal Mappings of the Disk. *Handbook of Complex Analysis: Geometric function theory, Volume 1*, 75–135, Edited by R. Kühnau, Elsevier Science B.V.

[44] Reich, E. and Strebel, K. (1974) Extremal quasiconformal mappings with given boundary values. *Contributions to analysis, a collection of papers dedicated to Lipman Bers*, Academic Press, New York, 375–391.

[45] Royden, H. (1971) Automorphisms and isometries of Teichmüller space in 'Advances of the Theory of Riemann Surfaces.' *Ann. Math. Stud.*, **66**, 369–384.

[46] Rudin, W. (1976) L^p-isometries and equimeasurability. *Indiana Univ. Math. J.*, **25**, no. 3, 215–228.

[47] Rudin, W. (1987) *Real and complex analysis.* McGraw-Hill.

[48] Shapiro, J. (1976) Mackey topologies, reproducing kernels, and diagonal maps on the Hardy and Bergman spaces. *Duke Math. J.*, **43**, 187–202.

[49] Strebel, K. (1961/2) Zur Frage der Eindeutigkeit extremaler quasikonforme Abbildungen des Einheitskreises. *Comment. Math. Helv.*, **36**, 306–323.

[50] Strebel, K. (1976) On the existence of extremal Teichmüller mappings. *J. Analyse Math.*, **30**, 464–480.

[51] Strebel, K. (1978) On quasiconformal mappings of open Riemann surfaces. *Comment. Math. Helv.*, **53**, 301–321.

[52] Strebel, K. (1984) *Quadratic differentials.* Springer-Verlag.

[53] Sullivan, D. and Thurston, W. (1986) Extending holomorphic motions. *Acta. Maths.*, **157**, 243–257.

[54] Teichmüller, O. (1939) Extremale quasikonforme Abbildungen und quadratische Differentiale. *Abh. Preuss. Akad. Wiss., math.-naturw. Kl.*, **22**, 1–197.

[55] Teichmüller, O. (1943) Bestimmung der extremalen quasikonformen Abbildungen bei geschlossenen orientierten Riemannschen Flächen. *Abh. Preuss. Akad. Wiss., math.-naturw. Kl.*, **4**, 1–42.

[56] Wolpert, S. (1982) The Fenchel–Nielsen deformation. *Ann. of Math. (2)* **115**, no. 3, 501–528.

[57] Yoa, G. (2004) Is there always an extremal Teichmüller mapping? *J. Anal. Math.*, **94**, 363–375.

Index

The manufacturer's authorised representative in the EU for product safety is Oxford University Press España S.A. of el Parque Empresarial San Fernando de Henares, Avenida de Castilla, 2 – 28830 Madrid (www.oup.es/en or product.safety@oup.com). OUP España S.A. also acts as importer into Spain of products made by the manufacturer.

www.ingramcontent.com/pod-product-compliance
Lightning Source LLC
LaVergne TN
LVHW011803200926
848484LV00003B/769

* 9 7 8 0 1 9 8 5 6 9 2 6 8 *